GCSE
Success

Combined Science

Higher Tier

Exam Practice Workbook

Tom Adams
Dan Evans
Dan Foulder

Biology

Cell Biology

Module 1: Structure of cells 6

Module 2: Organisation and differentiation 7

Module 3: Microscopy and microorganisms 8

Module 4: Cell division 9

Module 5: Metabolism – respiration 10

Module 6: Metabolism – enzymes 11

Transport Systems and Photosynthesis

Module 7: Cell transport 12

Module 8: Plant tissues, organs and systems 13

Module 9: Transport in plants 14

Module 10: Transport in humans 1 15

Module 11: Transport in humans 2 16

Module 12: Photosynthesis 17

Health, Disease and the Development of Medicines

Module 13: Non-communicable diseases 18

Module 14: Communicable diseases 19

Module 15: Human defences 20

Module 16: Fighting disease 21

Module 17: Discovery and development of drugs 22

Module 18: Plant diseases 23

Coordination and Control

Module 19: Homeostasis and negative feedback 24

Module 20: The nervous system 25

Module 21: The endocrine system 26

Module 22: Water and nitrogen balance 27

Module 23: Hormones in human reproduction 28

Module 24: Contraception and infertility 29

Inheritance, Variation and Evolution

Module 25: Sexual and asexual reproduction 30

Module 26: DNA 31

Module 27: The genetic code 32

Module 28: Inheritance and genetic disorders 33

Module 29: Variation and evolution 34

Module 30: Darwin and evolution 35

Module 31: Selective breeding and genetic engineering 36

Module 32: Classification 37

Contents

Ecosystems

Module 33: Organisms and ecosystems 38
Module 34: Studying ecosystems 39
Module 35: Feeding relationships 40
Module 36: Environmental change and biodiversity 41
Module 37: Recycling 43
Module 38: Farming and sustainability 44

Chemistry

Atomic Structure and the Periodic Table

Module 39: Atoms and the periodic table 45
Module 40: Atomic structure 46
Module 41: Electronic structure and the periodic table 47
Module 42: Groups 0, 1 and 7 48

Structure, Bonding and the Properties of Matter

Module 43: Chemical bonding 49
Module 44: Ionic and covalent structures 50
Module 45: States of matter: properties of compounds 51
Module 46: Metals, alloys and the structure and bonding of carbon 52

Quantitative Chemistry

Module 47: Mass and equations 53
Module 48: Moles, masses, empirical and molecular formula........................... 54
Module 49: Moles, solutions and equations 55

Chemical and Energy Changes

Module 50: Reactivity of metals and metal extraction 56
Module 51: Reactions of acids 57
Module 52: pH, neutralisation, acid strength and electrolysis 58
Module 53: Applications of electrolysis 59
Module 54: Energy changes in reactions 60

The Rate and Extent of Chemical Reactions

Module 55: Rates of reaction 61
Module 56: Collision theory, activation energy and catalysts 62
Module 57: Reversible reactions and equilibrium 63

Organic Chemistry

Module 58: Crude oil, hydrocarbons and alkanes 64

Module 59: Combustion and cracking of hydrocarbons 65

Chemical Analysis

Module 60: Purity, formulations and chromatography 66

Module 61: Identification of gases... 67

The Earth's Atmosphere and Resources

Module 62: Evolution of the atmosphere 68

Module 63: Climate change 69

Module 64: Atmospheric pollution.. 70

Module 65: Using the Earth's resources and obtaining potable water 71

Module 66: Alternative methods of extracting metals... 72

Module 67: Life-cycle assessment and recycling.. 73

Physics

Forces

Module 68: Forces.. 74

Module 69: Forces and elasticity... 75

Module 70: Speed and velocity.. 76

Module 71: Distance–time and velocity–time graphs... 77

Module 72: Newton's laws..78

Module 73: Forces, braking and momentum...79

Energy

Module 74: Changes in energy 80

Module 75: Conservation and dissipation of energy 81

Module 76: National and global energy resources...82

Waves

Module 77: Transverse and longitudinal waves 83

Module 78: Electromagnetic waves and properties 1 84

Module 79: Electromagnetic waves and properties 2 85

Electricity

Module 80: Circuits, charge and current 86
Module 81: Current, resistance and potential difference 87
Module 82: Series and parallel circuits 88
Module 83: Domestic uses and safety 89
Module 84: Energy transfers 90

Magnetism and Electromagnetism

Module 85: Permanent and induced magnetism, magnetic forces and fields 91
Module 86: Electromagnets and the motor effect 92

Particle Model of Matter

Module 87: The particle model and pressure 93
Module 88: Internal energy and change of state 94

Atomic Structure

Module 89: Atoms and isotopes 95
Module 90: Radioactive decay, nuclear radiation and nuclear equations 96
Module 91: Half-lives and the random nature of radioactive decay 97

Practice Exam Papers

Paper 1: Biology 1 98
Paper 2: Biology 2 108
Paper 3: Chemistry 1 120
Paper 4: Chemistry 2 131
Paper 5: Physics 1 143
Paper 6: Physics 2 155

Answers

169

The Periodic Table

200

1   This image shows some human liver cells, as seen through a very powerful light microscope.

(a) Which organelle is labelled X? .... (1)

(b) Unlike skin cells, these cells contain many mitochondria, but they cannot be seen in the image. Suggest why not.   (1)

....

(c) Why do the cells have many mitochondria?   (2)

....

....

(d) Liver cells have different features.
Arrange the features in order of size, starting with the largest.   (2)

gene	nucleus	cell	chromosome	cytoplasm

....

2   **(a)** A student is observing bacterial cells under the high power lens of a light microscope. She cannot see a nucleus in the cells and concludes that the cells do not contain DNA. Explain why this conclusion is wrong.   (1)

....

....

(b) Which of the following words best describes a bacterial cell? Tick (✓) **one** box.   (1)

Prokaryotic   ☐

Eukaryotic   ☐

Multicellular   ☐

Undifferentiated   ☐

For more help on this topic, see Letts GCSE Combined Science Higher Revision Guide pages 6–7

1 The diagram shows a fertilised egg cell (a zygote).

(a) The zygote is described as a **stem cell**. What does this term mean? (1)

....

(b) The zygote has a very different appearance to the root hair cell (pictured right). However, all cells have some structures in common. Write down **two** of these structures. (2)

1. ....

2. ....

2 **(a)** A student attends a school trip to a medical research laboratory, where he is given a talk from a scientist on the techniques and benefits of stem cell research.

Describe **two** applications of stem cell research that the scientist is likely to mention. (2)

1. ....

2. ....

(b) After the visit, one of the student's friends says she is opposed to stem cell research. Describe **one** objection that people have to stem cell research. (1)

....

....

3 A live sperm cell is observed under the microscope beating its tail. It moves across the field of view at a rate of 200 µm every 30 seconds.

Assuming the sperm travels at the same rate and in the same direction, how far will it have travelled in one hour? Give your answer in mm. (3)

....

For more help on this topic, see Letts GCSE Combined Science Higher Revision Guide pages 8–9

Organisation and differentiation

Module 2

1 (a) Daljit is looking at some cheek cells using a light microscope under low power. He decides that he wants a more magnified view. How should he adjust the microscope?

Draw a line from X to show which part of the microscope he should adjust.

X

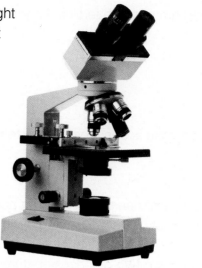

(1)

(b) Under higher power, Daljit clearly sees the nucleus, cytoplasm and cell membrane of the cheek cells. He would also like to see mitochondria and ribosomes. Give **two** reasons why he cannot see these structures.

(2)

....

....

(c) Using a specially fitted camera, Daljit takes a picture of the cells he sees. He measures the diameter of one cell on his photograph by drawing a line and using a ruler. The line is shown on the picture. It measures 3 cm.

If the microscope magnifies the image 400 times, calculate the actual size of the cell in μm.

You can use the following formula:

$$\text{magnification} = \frac{\text{size of image}}{\text{size of real object}}$$

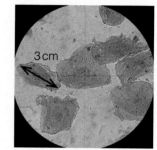

3 cm

Answer: .... μm   (3)

2 (a) Look at the photograph on the right. What type of microscope was used to take this picture?

(1)

....

(b) There are 27 bacterial cells in this picture. If the bacteria have reproduced from 1 cell every 20 minutes, how long have the bacteria been growing?

(2)

.... minutes

For more help on this topic, see Letts GCSE Combined Science Higher Revision Guide pages 10–11

1 The illustrations show a single-celled organism, called an amoeba, and a multicellular organism (a horse).

Not to scale

Explain why the horse has specialised organs in its breathing and digestive system, and the amoeba has none. (2)

....

....

....

....

2 Complete the following table, which compares the processes of mitosis and meiosis. (3)

Mitosis	Meiosis
Involved in asexual reproduction	....
....	Produces variation
Produces cells with 46 chromosomes	....

3 Jenny is studying chicken cells and is looking at some examples down the microscope. She draws these cells.

(a) Which type of cell division is shown here? .... (1)

(b) Give a reason for your answer to **(a)**. (1)

....

4 (a) Explain the difference between a **benign** and a **malignant** tumour. (2)

....

....

....

(b) Write down **two** lifestyle choices a person could make to reduce their chances of developing cancer. (2)

....

....

For more help on this topic, see Letts GCSE Combined Science Higher Revision Guide pages 12–13

Cell division

Module 4

HT **1** Isaac is running a marathon. Write a balanced symbol equation for the main type
of respiration that will be occurring in his muscles. (2)

....

2 Isaac's metabolic rate is monitored as part of his training schedule. He is rigged up to
a metabolic rate meter. This measures the volumes of gas that he breathes in and out.
The difference in these volumes represents oxygen consumption. This can be used in a
calculation to show metabolic rate.

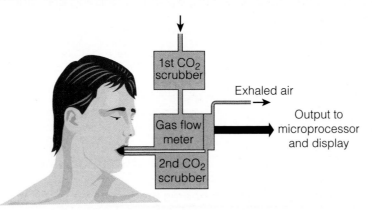

The table below shows some measurements taken from the meter over a period of one hour.

	Five minutes of jogging	Five minutes of rest	Five minutes of sprinting	Five minutes sprinting on an incline
Mean metabolic rate per ml oxygen used per kg per min	35	20	45	60

(a) The units of metabolic rate are expressed in the table as 'per kg'.
Why is this adjustment made? (2)

....

....

(b) Using the table, explain the difference in readings for jogging and sprinting. (2)

....

(c) Isaac does quite a lot of exercise. His friend Boris does not. How might Boris'
readings compare with Isaac's? Give a reason for your answer. (2)

....

....

For more help on this topic, see Letts GCSE Combined Science Higher Revision Guide pages 14–15

1 In an experiment to investigate the enzyme catalase, potato extract was added to a solution of hydrogen peroxide. The catalase in the potato catalysed the decomposition of the hydrogen peroxide and produced oxygen bubbles. The experiment was carried out at different temperatures and the results recorded in the table below.

Temperature / °C	0	10	20	30	40	50	60	70	80
Number of bubbles produced in one minute	0	10	24	40	48	38	8	0	0

(a) Plot a graph of these results on the graph paper. (3)

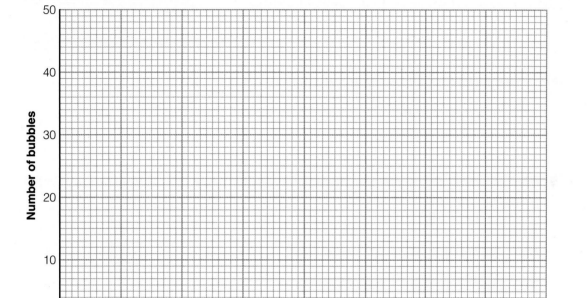

(b) Describe how the number of bubbles produced varies with the temperature of the reacting mixture. (2)

....

....

(c) Using the graph, estimate the optimum temperature for catalase to work at. (1)

....

(d) Factors other than temperature affect the activity of enzymes. Explain why the enzyme amylase, found in saliva, stops working when it gets to the stomach. (2)

....

....

For more help on this topic, see Letts GCSE Combined Science Higher Revision Guide pages 16–17

1 Cells rely on diffusion as a way of transporting materials inwards and outwards.

(a) Name **two** substances that move by diffusion **into** animal cells. (2)

.... and ....

(b) Name **one** substance that might diffuse **out** of an animal cell. .... (1)

2 Osmosis is a special case of diffusion involving water. Plants rely on osmosis for movement of materials around their various structures.

On the right is a diagram of three plant cells in the root of a plant. Cell **A** has a higher concentration of water than cell **C**.

(a) Explain how water can keep moving from cell **A** to cell **C**. .... (3)

....

....

(b) Which of the following are examples of osmosis? Tick (✓) the **three** correct options. (3)

Water evaporating from leaves ☐

Water moving from plant cell to plant cell ☐

Mixing pure water and sugar solution ☐

A pear losing water in a concentrated solution of sugar ☐

Water moving from blood plasma to body cells ☐

Sugar being absorbed from the intestine into the blood ☐

3 Rabia is investigating how plant cells respond to being surrounded by different concentrations of sugar solution. She places some rhubarb tissue into pure water and then observes the cells under the microscope.

(a) Describe and explain the appearance of the rhubarb cells. (2)

....

....

(b) Rabia then puts some rhubarb tissue into a strong salt solution. Describe how the cells would change in appearance if she observed them under the microscope. (2)

....

....

For more help on this topic, see Letts GCSE Combined Science Higher Revision Guide pages 20–21

1 This diagram shows a magnified view of the inside of a leaf. Complete the missing labels. (3)

Waxy cuticle

Palisade cells

Stoma

(a) ....

(b) ....

(c) ....

2 Plants all share a basic structure consisting of four main organs.

Describe the function that each organ performs in the plant. (4)

Roots: ....

Stem: ....

Leaf: ....

Flower: ....

3 Describe **three** adaptations of xylem vessels that make them suited to the job they do. (3)

....

....

4 The diagram shows xylem and phloem tissue.

(a) State **one** structural difference between the two tissues. (1)

....

....

Xylem   Phloem

Xylem   Phloem

(b) Small, herbivorous insects called aphids are found on plant stems. They have piercing mouthparts that can penetrate down to the phloem.

Explain the reasons for this behaviour. (2)

....

....

For more help on this topic, see Letts GCSE Combined Science Higher Revision Guide pages 22–23

Plant tissues, organs and systems

Module 8

1 In an experiment, a plant biologist carried out an investigation to measure the rate of transpiration in a privet shoot. She set up three tubes like the one in the diagram, measured their mass and exposed them each to different conditions.

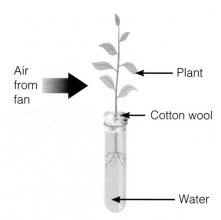

- **A** – Left to stand in a rack

- **B** – Cold moving air from a fan was blown over it

- **C** – A radiant heater was placed next to it

The tubes were left for six hours and then their masses were re-measured. The biologist recorded the masses in this table.

Tube	A	B	C
Mass at start (g)	41	43	45
Mass after six hours (g)	39	35	37
Mass loss (g)	2	8	5
% mass loss	4.9		11.9

(a) Calculate the percentage mass loss in tube B. Show your working. (2)

....

....

(b) Which factor increased the rate of transpiration the most? .... (1)

(c) Evaporation from the leaves has increased in tubes B and C. Describe how this would affect water in the xylem vessels of the plant. (1)

....

2 Guard cells respond to light intensity by opening and closing stomata. Explain how this occurs. In your answer, use ideas about osmosis and turgidity. (6)

....

....

....

....

....

....

Continue your answer for this question on a separate piece of paper.

For more help on this topic, see Letts GCSE Combined Science Higher Revision Guide pages 24–25

1 The graph shows the number of people who die from coronary heart disease per 100 000 people in different parts of the UK.

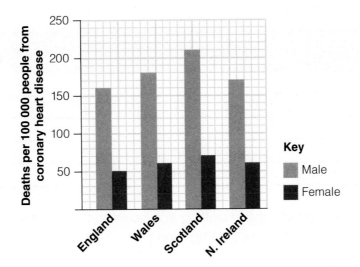

(a) (i) Which country has the highest number of deaths per 100 000? (1)

....

(ii) Calculate the difference between this country's death rate for males and the country with the smallest death rate for males. (1)

....

(b) Describe the pattern between death rates from coronary heart disease in men and women. (1)

....

2 The diagram shows the human circulatory system.

Match the numbers on the diagram with the words listed below. Write the appropriate numbers in the boxes. (4)

Artery ☐

Capillaries in the body ☐

Vein ☐

Capillaries in the lungs ☐

Deoxygenated blood ■   Oxygenated blood ■

For more help on this topic, see Letts GCSE Combined Science Higher Revision Guide pages 26–27

1 The table shows some data from an experiment measuring the effect of different doses of warfarin in two samples of rat blood. Warfarin prevents blood from clotting, killing the rat as a result of internal bleeding.

Warfarin dose / mg per kg body weight	Time for clot to form in sample 1	Time for clot to form in sample 2
0.2	52	61
0.4	101	99
0.6	193	197
0.8	300	311
1	978	984

(a) Calculate the average change in clotting rate between 0.2 and 1.0 mg of warfarin per kilogram of body weight. Show your working. (2)

....

....

(b) Explain why the warfarin dose was measured in mg per kg of body weight. (1)

....

(c) Describe the pattern of results shown in the data. (1)

....

2 Scientists are studying the performance of pearl divers living on a Japanese island. They have timed how long they can stay underwater. The scientists have also measured recovery time for the divers' breathing rates after a dive. The data is shown in the table below.

Vital capacity / litres	Max. time under water / mins
3.5	2.5
4.0	2.7
4.3	2.8
4.5	2.9
4.6	3.0

Vital capacity is the maximum volume of air that the lungs can hold at any one time.

One of the scientists suggests that having a larger vital capacity allows a diver to stay underwater for longer. Do you agree with her? Give a reason for your answer. (2)

Agree / disagree: ....

Reason: ....

....

For more help on this topic, see Letts GCSE Combined Science Higher Revision Guide pages 28–29

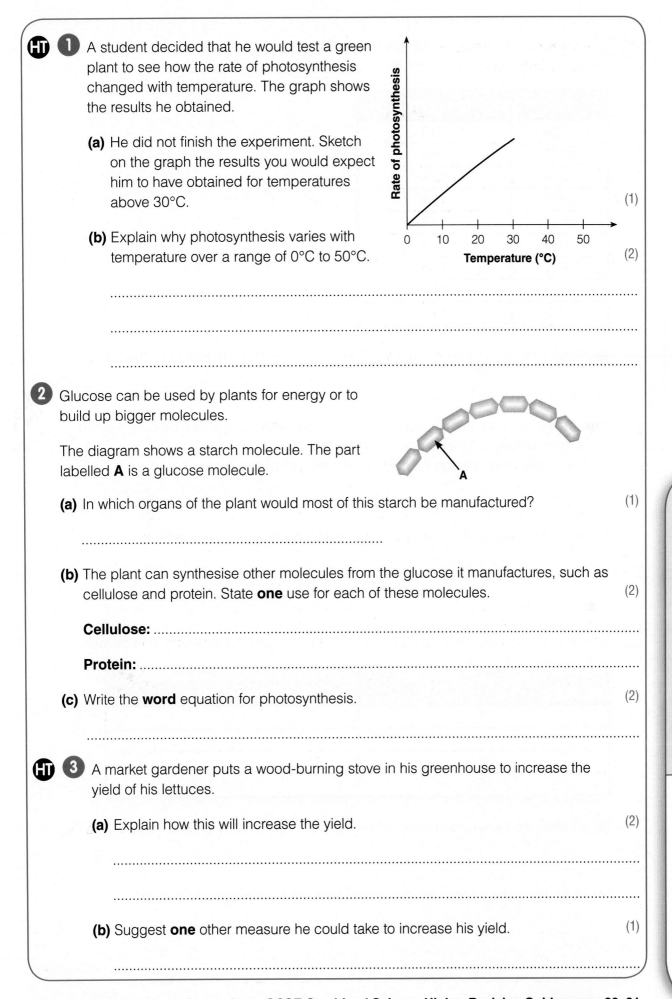

HT **1** A student decided that he would test a green plant to see how the rate of photosynthesis changed with temperature. The graph shows the results he obtained.

Rate of photosynthesis

Temperature (°C)

0   10   20   30   40   50

(a) He did not finish the experiment. Sketch on the graph the results you would expect him to have obtained for temperatures above 30°C.

(1)

(b) Explain why photosynthesis varies with temperature over a range of 0°C to 50°C.

(2)

....

....

....

2 Glucose can be used by plants for energy or to build up bigger molecules.

The diagram shows a starch molecule. The part labelled **A** is a glucose molecule.

A

(a) In which organs of the plant would most of this starch be manufactured?

(1)

....

(b) The plant can synthesise other molecules from the glucose it manufactures, such as cellulose and protein. State **one** use for each of these molecules.

(2)

Cellulose: ....

Protein: ....

(c) Write the **word** equation for photosynthesis.

(2)

....

HT **3** A market gardener puts a wood-burning stove in his greenhouse to increase the yield of his lettuces.

(a) Explain how this will increase the yield.

(2)

....

....

(b) Suggest **one** other measure he could take to increase his yield.

(1)

....

For more help on this topic, see Letts GCSE Combined Science Higher Revision Guide pages 30–31

1 Obesity is a non-communicable condition. The table shows how obesity in a population of children aged between two and ten changed between 2005–2013.

Year	% of obese children
2005	9.9
2006	10.2
2007	10.8
2008	11.3
2009	11.8
2010	12.3
2011	12.8
2012	13.3
2013	13.7

(a) Predict the percentage of obese children in the UK for 2014 based on this trend. (1)

.............................%

(b) If the body's daily energy requirements are exceeded, sugar can be converted to storage products; for example, fat under the skin.
Name **one** other storage product and where it would be found. (2)

....

(c) Write down **two** other conditions / diseases that are non-communicable. (2)

....

....

2 The table shows how smoking can affect a person's chances of getting lung cancer.

Number of cigarettes smoked per day	Increased chance of lung cancer compared to non-smokers
5	×4
10	×8
15	×12
20	×16

(a) Estimate the increased chance of lung cancer if someone smoked 25 cigarettes a day. (1)

....

(b) Write down **one** harmful effect that smoking can have on unborn babies. (1)

....

....

For more help on this topic, see Letts GCSE Combined Science Higher Revision Guide pages 34–35

1 Microorganisms consist of bacteria, viruses, fungi and protists. Many cause harm to the human body.

(a) Write down the term that describes these disease-causing organisms. (1)

....

(b) Harmful microorganisms produce symptoms when they reproduce in large numbers. Write down **two** ways in which microorganisms cause these symptoms. (2)

1. ....

2. ....

2 The picture below shows the bacterium that causes cholera.

(a) Write down **two** symptoms of cholera. (2)

1. ....

2. ....

(b) Explain why cholera spreads rapidly in natural disaster zones. (2)

1. ....

2. ....

3 Malaria kills many thousands of people every year. The disease is common in areas that have warm temperatures and stagnant water.

(a) Explain why this is. (2)

....

....

(b) A protist called *Plasmodium* lives in the salivary glands of the female *Anopheles* mosquito.

From the box below, choose a word that describes each organism. (2)

parasite	disease	symptom	vector	consumer	host

Mosquito: ....   **Plasmodium:** ....

(c) Samit, an African villager, believes that having mosquito nets around the beds of family members and taking antiviral remedies will reduce their risk of catching malaria.

Explain why he is only partially correct. (2)

....

....

Communicable diseases

Module 14

For more help on this topic, see Letts GCSE Combined Science Higher Revision Guide pages 36–37

1 The diagram shows a white blood cell producing small proteins as part of the body's immune system.

(a) What is the name of these proteins?

.... (1)

(b) These proteins will eventually lock on to specific invading pathogens. Describe what happens next to disable the pathogens. (1)

....

....

(c) Below are the names of some defence mechanisms that the body uses. Match each defence mechanism with the correct function. The first one has been done for you. (3)

Epithelial cells in respiratory passages		engulf pathogens.
Phagocytes		contain enzymes called lysozymes that break down pathogen cells.
Tears		contains acid to break down pathogen cells.
Stomach		trap pathogens in mucus.

2 The graph shows the antibody levels in Dominic after he contracted the flu. The flu pathogen first entered his body two days before point X. There was then a second invasion at point Y.

(a) Name **one** transmission method by which Dominic could have caught the flu virus. (1)

....

(b) After how many days did the antibodies reach their maximum level? (1)

.... days

(c) What is the difference in antibody level between point Y and this maximum? Show your working. (2)

.... arbitrary units

(d) Explain, using your knowledge of memory cells, the difference between these two levels. (2)

....

....

For more help on this topic, see Letts GCSE Combined Science Higher Revision Guide pages 38–39

HT **1** The photograph shows the structure of the human immunodeficiency virus (HIV). For decades it has spread throughout the world, especially in developing countries.

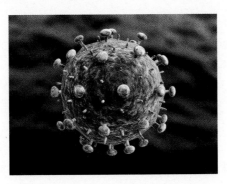

A vaccine is now being developed that shows promising results. It works by mimicing the shapes and structures of HIV proteins. Scientists hope the immune system may be 'educated' to attack the real virus. A specially designed adenovirus shell can protect the vaccine genes until they are in a cell that can produce the vaccine protein.

Using your knowledge of the immune response and immunological memory, describe and explain how antibodies can be produced against the HIV virus. (6)

....

....

....

....

....

....

....

....

....

2 Explain why antibiotics are becoming increasingly less effective against 'superbugs' such as MRSA. (3)

....

....

....

....

Fighting disease

Module 16

For more help on this topic, see Letts GCSE Combined Science Higher Revision Guide pages 40–41

1 A pharmaceutical company is carrying out a clinical trial on a new drug called alketronol. They are testing it to see whether it produces significant adverse (harmful) events in a sample of 226 patients.

(a) Apart from checking for adverse events, write down **two** other reasons that a company carries out clinical trials. (2)

1. ....

2. ....

(b) The kind of trial carried out is a double blind trial. What does this term mean? (2)

....

....

....

(c) Data from the trial is shown in the table below.

Adverse event	Alketronol	Placebo
	Number of patients	**Number of patients**
Pain	4	3
Cardiovascular	21	20
Dyspepsia	7	6
Rash	10	1

(i) Calculate the percentage of patients **in the trial** who suffered a cardiovascular event while taking alketronol. Show your working. (2)

.... %

(ii) A scientist is worried that alketronol may trigger heart attacks. Is there evidence in the data to support this view? Explain your answer. (2)

....

....

....

(iii) Which other adverse event might cause concern? Give a reason for your decision. (2)

....

....

For more help on this topic, see Letts GCSE Combined Science Higher Revision Guide pages 42–43

1 Ash dieback, or *Chalara*, is caused by a fungus called *Hymenoscyphus fraxineus*. *Chalara* results in loss of leaves, crown dieback and bark damage in ash trees. Once a tree is infected, the disease is usually fatal because the tree is weakened and becomes prone to pests or pathogens.

The map gives an indication of where cases of *Chalara* were reported in 2012 in the UK.

Scientists have also discovered that:

- *Chalara* spores are unlikely to survive for more than a few days

- spores can be dispersed by winds blowing from mainland Europe

- trees need a high dose of spores to become infected

- there is a low probability of dispersal on clothing or animals and birds.

Key:
■ = infection confirmed

(a) Damage to leaves can be caused by lack of certain minerals that plants need.

Write down **one** mineral deficiency and how it can affect leaves.

Mineral deficiency: .... **Leaf appearance:** .... (2)

(b) Which **one** of the following conclusions is supported by evidence from the map? Tick (✓) **one** box. (1)

Chalara is limited to the East of England. ☐

Spores of *Chalara* arrived in England by being carried on winds from Europe. ☐

There is a high concentration of *Chalara* cases in the East of England. ☐

Ash trees in north-west Scotland are resistant to *Chalara*. ☐

(c) One scientist suggests that cutting down and burning infected trees could eradicate the disease.

(i) Explain how this method could be effective. (2)

....

....

(ii) Give **one** reason why this control method may not stop the spread of *Chalara*. (1)

....

For more help on this topic, see Letts GCSE Combined Science Higher Revision Guide pages 44–45

1 From the box below, choose three words to complete the information about how conditions are kept stable in the human body.　(3)

effectors	spine	receptors	homeostasis	hormones	glands

Certain factors have to be kept constant in the body. This is achieved by a process called .... In order for this to happen, the central nervous system (CNS) needs to receive information from the environment. This is accomplished through .... such as light-sensitive cells on the retina. Once the information has been relayed, the CNS brings about appropriate changes through ....

HT **2** This diagram shows how production of the hormone adrenaline is controlled.

Hypothalamus → **CRH** Corticotropin-releasing hormone → Pituitary → **ACTH** Adrenocorticotropic hormone → Adrenal glands → **Adrenaline**

(a) What name is given to this process, where a system resists a change from a norm (set point) level? ....　(1)

(b) Another hormone, cortisol, is produced by the adrenal glands. Its production is also governed by CRH and ACTH production, in the same way as adrenaline. Cortisol increases nutrient distribution, reduces inflammation, and also takes part in water control. In Addison's disease, the adrenal glands fail to produce enough cortisol.

(i) What is the effect of Addison's disease on the production of ACTH?　(1)

....

(ii) Using the information above, give **one** symptom of Addison's disease.　(1)

....

(iii) Here is some data taken from adult blood samples.

Patient	A	B	C	D	E
Cortisol in blood / µg per litre	31.0	19.2	20.5	1.2	16.0

Which patient is most likely to have Addison's disease? ....　(1)

(iv) A doctor injects some cortisol into this patient's blood, then takes another sample. The reading is now 7 µg per litre. If we assume that the patient has 5 litres of blood in their body, calculate the amount of cortisol in this person's blood. Show your working.　(2)

....

For more help on this topic, see Letts GCSE Combined Science Higher Revision Guide pages 48–49

1 Complete the missing labels in this diagram of a motor neurone. (2)

(a) ....

Axon

Effector / muscle

(b) ....

2 A company is developing technology that allows robots to make decisions in a form of artificial intelligence. As part of their research, the company tries to model the human nervous system.

(a) Write down the name given to the part of the nervous system that processes information from the senses. .... (1)

(b) Describe the other parts of the human nervous system that would need to be reproduced so that the robots could detect stimuli in the environment. In your answer, use the words in the box below. (3)

receptor	sensory neurone	brain

....

....

....

3 The diagram shows a junction (**X**) between two nerve cells.

Impulse

X

Impulse

(a) What name is given to this nerve junction? .... (1)

(b) In order for the impulse to be transmitted from one neurone to the next, a transmitter substance needs to be released. Describe the sequence of events involved. (3)

....

....

....

For more help on this topic, see Letts GCSE Combined Science Higher Revision Guide pages 50–51

The nervous system

Module 20

1 Label the gland shown on the diagram and add the name of a hormone it produces. (2)

| Gland: |
| Hormone: |

2 A new nanotechnology device has been developed for people with diabetes – it can detect levels of glucose in the blood and communicate this information to a hormone implant elsewhere in the body. The implant releases a precise quantity of hormone into the bloodstream when required.

(a) Explain how this device could help a person with type 1 diabetes who has just eaten a meal. (2)

....

....

(b) Explain why a person with type 2 diabetes might not have as much use for this technology. (2)

....

....

3 Complete the missing information in the table, which is about different endocrine glands in the body. (4)

Gland	Hormones produced
Pituitary gland	.... and ....
Pancreas	Insulin and glucagon
....	Thyroxine
....	Adrenaline
Ovary	.... and ....
Testes	Testosterone

For more help on this topic, see Letts GCSE Combined Science Higher Revision Guide pages 52–53

1 A marathon runner is resting the day before she competes in a race. The table shows the water that she gains and loses during the day.

Gained	Water gained (ml)	Lost	Water lost (ml)
In food	1000	In urine	
From respiration	300	In sweat	800
Drinking	1200	In faeces	100
Total gained		**Total lost**	2500

(a) How much water does the runner lose in urine during the day? .... (1)

(b) What can you say about the total water gained and the total water lost in the day? Why is this important? (2)

....

....

(c) The runner runs a marathon the next day. Suggest and explain how the figures shown in the table may alter during the day of the race. (4)

....

....

....

2 The diagram shows a nephron.

Your responses to the following questions should be **A**, **B** or **C**.

(a) Where does selective reabsorption occur? (1)

....

(b) Where does salt regulation occur? .... (1)

(c) In which region does filtration occur? .... (1)

(d) Explain how the brain and kidneys work together to restore water levels in the blood when the body is dehydrated. (4)

....

....

....

For more help on this topic, see Letts GCSE Combined Science Higher Revision Guide pages 54–55

Water and nitrogen balance

Module 22

1 The graph shows the thickness of the uterus during the menstrual cycle. Use the graph and your scientific knowledge to explain what happens in the woman's ovaries and uterus between days 5 and 28. (3)

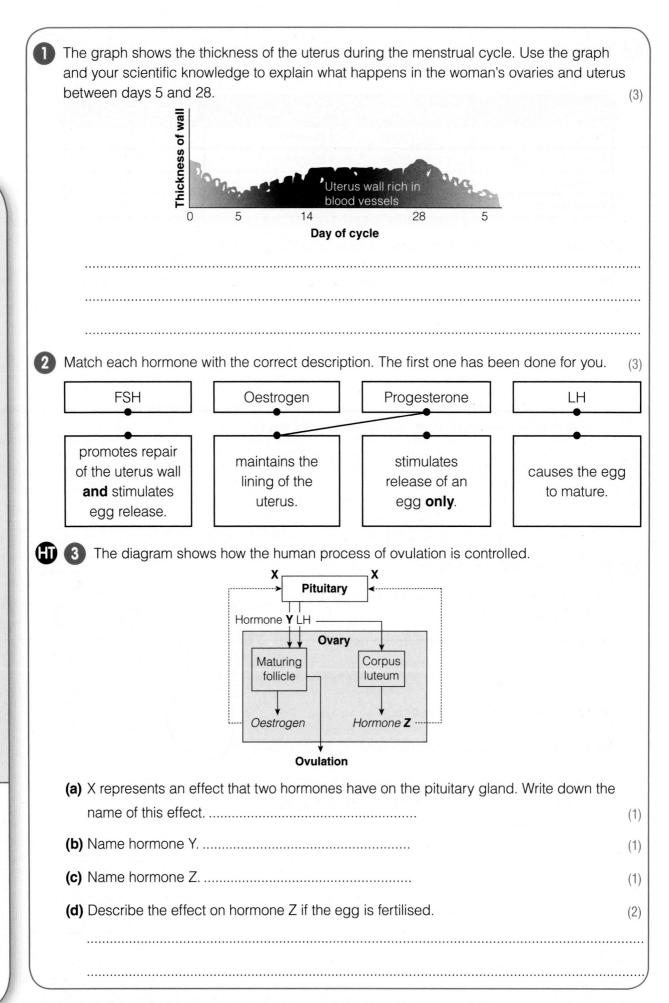

Day of cycle

....

....

....

2 Match each hormone with the correct description. The first one has been done for you. (3)

FSH	Oestrogen	Progesterone	LH

| promotes repair of the uterus wall **and** stimulates egg release. | maintains the lining of the uterus. | stimulates release of an egg **only**. | causes the egg to mature. |

HT **3** The diagram shows how the human process of ovulation is controlled.

(a) X represents an effect that two hormones have on the pituitary gland. Write down the name of this effect. .... (1)

(b) Name hormone Y. .... (1)

(c) Name hormone Z. .... (1)

(d) Describe the effect on hormone Z if the egg is fertilised. (2)

....

....

For more help on this topic, see Letts GCSE Combined Science Higher Revision Guide pages 56–57

Tim and Margaret are finding it hard to conceive a child. They visit a fertility clinic and meet some other couples. The table shows some information about the problem that each couple has.

Couple	Problem causing infertility	Percentage of infertile couples with this problem	Percentage success rate of treatment
Tim and Margaret	Blocked fallopian tubes	13	20
Rohit and Saleema	Irregular ovulation	16	75
Leroy and Jane	No ovulation	7	95
Gary and Charlotte	Low sperm production	15	10
Ian and Kaye	No sperm production	21	10
Stuart and Mai	Unknown cause	28	–

(a) Which couple has the best chance of being successfully treated? (1)

....

(b) In how many of the six couples is the problem known to be with the female? (1)

....

(c) The treatment of irregular ovulation and no ovulation have the highest success rates. Explain why treating irregular ovulation would produce more pregnancies in the whole population. (2)

....

(d) Leroy and Jane are considering two methods to help them have children. The first is to have an egg donated by another woman. The second is to arrange for another woman to conceive the child using sperm from Leroy, then give birth to it (surrogacy). What are the advantages and disadvantages of each method? (4)

....

....

....

....

....

2 Explain how the contraceptive pill works. In your answer, name any hormones involved. (2)

....

....

....

For more help on this topic, see Letts GCSE Combined Science Higher Revision Guide pages 58–59

1 From the box below, choose **three** words to complete these sentences. (3)

| zygotes | gametes | diploid | haploid | mitosis | meiosis |

Eggs and sperm are .... They are .... because they contain one set of chromosomes. Eggs and sperm are produced in the ovaries and testes by ....

2 Tick (✓) the **two** statements about causes of variation that are true. (2)

Meiosis shuffles genes, which makes each gamete unique. ☐

Gametes fuse randomly. ☐

Zygotes fuse randomly. ☐

Mitosis shuffles genes, which makes each gamete the same. ☐

3

Sexual reproduction is the best strategy for organisms because it allows variation and therefore greater adaptation.

Asexual reproduction is better because when an organism is well adapted, it can produce exact copies of itself.

(a) John and Ayesha disagree about which type of reproduction is most beneficial to organisms. State which explanation, if any, is correct. Give the reasons for your choice. (3)

....

....

....

....

(b) Apart from variation, write down **one** other difference between sexual and asexual reproduction. (1)

....

....

(c) Describe how yeast carries out asexual reproduction. (2)

....

....

For more help on this topic, see Letts GCSE Combined Science Higher Revision Guide pages 62–63

1 Tick (✓) the statements about the Human Genome Project (HGP) that are true. (3)

The genome of an organism is the entire genetic material present in its adult body cells. ☐

The data produced from the HGP produced a listing of amino acid sequences. ☐

The HGP involved collaboration between US and UK geneticists. ☐

The project allowed genetic abnormalities to be tracked between generations. ☐

The project was controversial as it relied on embryonic stem cells. ☐

2 Studies of genomes can help scientists work out the evolutionary history of organisms by comparing the similarity of particular DNA sequences that code for a specific protein.

The table shows the percentage DNA coding similarity for protein A in different organisms.

Species	% DNA coding similarity between species and humans for protein A
Human	100
Chimpanzee	100
Horse	88.5
Fish	78.6
Yeast	67.3
Protist	56.6

(a) What evidence is there in the table that closely related organisms developed from a recent common ancestor? (1)

....

(b) Using only the information from the table, which invertebrate is the most closely related to humans? .... (1)

3 The Human Genome Project has enabled specific genes to be identified that increase the risk of developing cancer in later life. Two of these genes are the *BRCA1* and *BRCA2* mutations that increase the risk of developing breast cancer.

(a) If women are prepared to take a genetic test, how could this information help doctors advise women about breast cancer? (2)

....

....

(b) If a woman possesses these mutations, it does not mean that she will definitely develop breast cancer. Why is this? (2)

....

....

DNA

Module 26

For more help on this topic, see Letts GCSE Combined Science Higher Revision Guide pages 64–65

1 The molecule DNA is a double helix made of two complementary strands.

(a) Write down the bases that pair with **T** and **C**. (1)

T pairs with ....

C pairs with ....

(b) How many bases code for **one** amino acid when a protein molecule is made? (1)

....

HT **2** Mutations occur when genes on DNA cause them to code for different proteins (or sequences of amino acids).

(a) State **two** causes of mutation. (2)

1. ....

2. ....

(b) A change occurs in a section of DNA that leads to a new protein being formed.

Explain how this is possible and why the protein is not able to perform its function. (3)

....

....

....

HT **3** DNA is constructed from individual building blocks called **nucleotides**.

Describe the structure of a nucleotide and how the sequence of these building blocks codes for amino acids. (4)

....

....

....

....

....

For more help on this topic, see Letts GCSE Combined Science Higher Revision Guide pages 66–67

1 Raj is the owner of two dogs, both of which are about two years old. Both dogs are black in colour and came from the same litter of puppies.

(a) A dog's adult body cell contains 78 chromosomes. How many chromosomes would be in a male dog's sperm cells? (1)

....

(b) The dogs' mother had white fur and the father had black fur. Using what you know about dominant genes, suggest why there were no white puppies in the litter. (2)

....

....

HT **(c)** One year later, one of the black puppies mated with a white-haired dog. She had four puppies. Two had black fur and two had white fur. The letters **B** and **b** represent the alleles for fur colour: **B** for black fur and **b** for white fur.

Draw a fully labelled genetic diagram to explain this. Show which offspring would be black and which would be white. (3)

2 Complete these two different crosses between a brown-eyed parent and a blue-eyed parent. (4)

(a)   Brown eyes × Blue eyes

Parents   (BB)   ×   (bb)

Gametes   ◯ ◯   ◯ ◯

Offspring   ◯ ◯   ◯ ◯

Phenotype [ ] [ ] [ ] [ ]

(b)   Brown eyes × Blue eyes

Parents   (Bb)   ×   (bb)

Gametes   ◯ ◯   ◯ ◯

Offspring   ◯ ◯   ◯ ◯

Phenotype [ ] [ ] [ ] [ ]

For more help on this topic, see Letts GCSE Combined Science Higher Revision Guide pages 68–69

1 This is an evolutionary tree for some of our present-day vertebrates. Where possible, use the diagram to answer these questions.

(a) How many millions of years ago did the testudina appear? (1)

....

(b) (i) In what geological time period did the dinosaurs become extinct? (1)

....

(ii) How do scientists know that dinosaurs once lived on Earth? (1)

....

(c) What group of animals alive today is most closely related to the snake? (1)

....

(d) Which ancestor is shared by dinosaurs, crocodiles and the giant lizard, but is not an ancestor of tortoises? (1)

....

For more help on this topic, see Letts GCSE Combined Science Higher Revision Guide pages 70–71

1 Peppered moths are usually pale and speckled. They are often found amongst the lichens on silver birch tree bark. The data below estimates the average number of peppered moths spotted in a city centre before and after the Industrial Revolution.

Month	Before Industrial Revolution		After Industrial Revolution	
	Pale	**Dark**	**Pale**	**Dark**
June	1261	102	87	1035
July	1247	126	108	1336
August	1272	93	72	1019

(a) Complete the table by calculating the mean number of each colour of moth during the summer months. (2)

Before Industrial Revolution		After Industrial Revolution	
Pale	**Dark**	**Pale**	**Dark**
....	....	....	....

(b) Draw a bar graph to represent your results. (2)

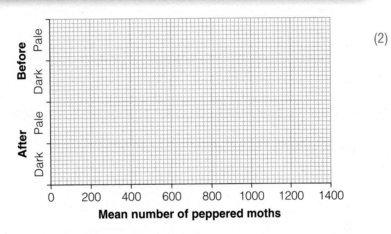

(c) Why do you think there were more pale-coloured moths than dark-coloured moths before the Industrial Revolution? (1)

....

(d) Explain why the number of dark-coloured peppered moths increased significantly during the Industrial Revolution. (2)

....

....

....

2 Lucy is an early hominid fossil that is 3.2 million years old. Why was this a significant find? (1)

....

For more help on this topic, see Letts GCSE Combined Science Higher Revision Guide pages 72–73

Darwin and evolution

Module 30

1 (a) A sheep farmer wants to breed sheep that grow high quality wool. What four stages of selective breeding can he use to produce his desired variety?   (3)

The first stage has been done for you.

1. Choose males and females that produce good quality wool.

2. ....

3. ....

4. ....

(b) Apart from high quality wool, name another characteristic that a farmer might want to selectively breed into his flock.   (1)

....

2 Describe how genetic engineering is different from selective breeding.   (2)

....

....

....

3 Explain the benefit of each of these examples of genetic engineering.

(a) Resistance to herbicide in soya plants   (1)

....

....

(b) Inserting beta-carotene genes into rice plants   (1)

....

....

4 Some people think that genetically engineering resistance to herbicides in plants could have unforeseen consequences. Give **one** example of a harmful effect.   (1)

....

....

....

For more help on this topic, see Letts GCSE Combined Science Higher Revision Guide pages 74–75

1 Lions, tigers and leopards are all carnivorous big cats. They have five toes on their front paws and four toes on their back paws. Their claws can be drawn back to avoid damage. They all roar. Tigers and leopards tend to be solitary animals but lions live in prides of females with one dominant male.

(a) Underline **one** piece of evidence from the information above that suggests that lions, tigers and leopards are all descended from a common ancestor. (1)

(b) This table shows how some scientists have named four species of large cat.

	Genus	Species
Lion	Panthera	leo
Tiger	Panthera	tigris
Leopard	Panthera	pardus
Snow leopard	Uncia	uncia

Are leopards more closely related to tigers or snow leopards? Explain your answer. (2)

....

....

....

2 Archaeopteryx is an ancient fossilised species of bird. When first discovered, scientists found it hard to classify.

Using features shown in the picture, explain why Archaeopteryx is difficult to classify. (2)

....

....

....

For more help on this topic, see Letts GCSE Combined Science Higher Revision Guide pages 76–77

1 Complete the following passage about adaptations using words from the box below. (3)

environment	population	features	community	characteristics
survival	evolutionary	predatory	suited	

Adaptations are special .... or .... that make a living

organism particularly well .... to its .... Adaptations

are part of an .... process that increases a living organism's chance of

....

2 A new species of insectivorous mammal has been discovered in Borneo. It has been observed in rainforest undergrowth and more open savannah-like areas. Scientists have called the creature a *long-nosed batink*. They have studied its diet and obtained this data.

Food	Ants	Termites	Aphids	Beetles	Maggots	Bugs	Grubs
Mass eaten per day / g	275	380	320	75	150	20	110

(a) Plot the data for termites, beetles, bugs and grubs as a bar chart. (3)

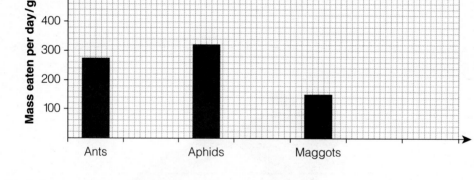

(b) Calculate the percentage of the batink's diet that is made up from termites. Show your working. (2)

.... %

(c) From what you know about the batink, suggest **two** behavioural adaptations it might have that makes it successful. (2)

....

....

For more help on this topic, see Letts GCSE Combined Science Higher Revision Guide pages 80–81

1 A meadow supports a wide variety of animals and plants. George is carrying out a survey of the meadow to assess the populations of organisms found there.

(a) State the term that describes the meadow as a place for organisms to live.   (1)

....

(b) Which word describes the different populations in the meadow and their interaction with the physical factors found there?   (1)

....

(c) George has laid pitfall traps in the meadow to capture and count soil invertebrates. He notices that there are many flying insects that are too difficult to count and identify.

Suggest an item of apparatus he could use to survey the flying insects.   (1)

....

(d) George uses a 0.25 m² quadrat to survey the plant populations. He lays ten quadrats in one corner of the field and finds a mean count of 16 meadow buttercups per quadrat. He estimates the area of the meadow to be 5000 m².

Calculate the expected number of buttercups in the whole meadow.
Show your working.   (2)

....

....

(e) George finds these two invertebrates in his pitfall traps.

Beetle   Snail

(i) The beetle feeds off other insects. Explain how a decrease in the number of beetles will eventually result in their numbers rising.   (2)

....

....

(ii) Thrushes eat snails and worms. Describe what would happen to the number of snails if large numbers of thrushes arrived in their habitat.   (1)

....

For more help on this topic, see Letts GCSE Combined Science Higher Revision Guide pages 82–83

Studying ecosystems   Module 34

1 Apple trees are grown in orchards in temperate climates. They are part of a wider food web.

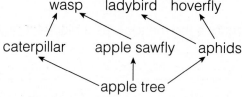

(a) The apple tree is a **producer**. What is meant by this term? (1)

....

(b) What is the source of energy for this food web? .... (1)

(c) Name a secondary consumer in the food web. .... (1)

2 Bird populations are a good indicator of environmental sustainability and allow scientists to track environmental changes in particular habitats.

Scientists measured the numbers of farmland birds and woodland birds in the UK between 1972 and 2002. The results are show below.

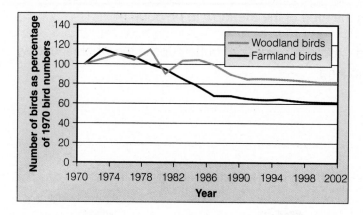

(a) Explain clearly how the numbers of farmland birds have changed between 1972 and 2002. (4)

....

....

....

....

(b) Suggest a reason for the overall change in numbers of farmland birds. (1)

....

(c) The government wants to reverse these changes by 2020. Suggest **one** thing it could do that would help to achieve this. (1)

....

For more help on this topic, see Letts GCSE Combined Science Higher Revision Guide pages 84–85

1 Circle the renewable resources. (1)

     water   **minerals**   **oil**   **wind**   **coal**   **wood**

2 Explain how rising average global temperatures may have an effect on the Earth.
Use the headings below to structure your answer. (3)

Climate zones around the world:

....

....

Sea levels:

....

....

Ice caps and glaciers:

....

....

3 At the moment, the human population is increasing exponentially.

(a) Sketch a graph that shows this increase. (2)

(b) On the *y*-axis, add a suitable unit for the population. (1)

(c) Suggest **two** reasons for this 'population explosion'. (2)

....

....

For more help on this topic, see Letts GCSE Combined Science Revision Guide page 86

1. In Ireland, four species of bumble bee are now endangered. Scientists are worried that numbers may become so low that they are inadequate to provide pollination to certain plants.

State **two** reasons why some organisms become endangered. (2)

....

....

2. **(a)** What is deforestation? Tick (✓) the correct definition. (1)

Planting new trees ☐

Forest fires caused by hot weather ☐

Cutting down large areas of forest ☐

Polluting national parks with litter ☐

(b) Which of the following is a consequence of deforestation? Tick (✓) the correct answer. (1)

Decrease in soil erosion in tropical regions ☐

Increase in atmospheric carbon dioxide ☐

Increase in average rainfall ☐

Increase in habitat area ☐

3. Other than using wood for timber, give **two** other reasons for large-scale deforestation. (2)

....

....

4. Circle the correct options in the sentences below. (6)

When deforestation occurs in **tropical / arctic / desert** regions, it has a devastating impact on the environment.

The loss of **trees / animals / insects** means less photosynthesis takes place, so less **oxygen / nitrogen / carbon dioxide** is removed from the atmosphere.

It also leads to a reduction in **variation / biodiversity / mutation**, because some species may become **protected / damaged / extinct** and **habitats / land / farms** are destroyed.

5. Sometimes when land has been cleared of forests to grow crops, farmers stop producing good yields after a few years. Explain why. (1)

....

....

For more help on this topic, see Letts GCSE Combined Science Revision Guide page 87

1 Choose the correct words from the list to complete this paragraph on the water cycle: (3)

| boiled | precipitation | rivers | organisms | evaporated | condensation |

The water cycle provides fresh water for .... on land before draining into the seas. Water is .... from seawater to form vapour in the atmosphere. This vapour then forms clouds and falls as ....

2 A group of students wanted to investigate factors affecting decay. They mixed soil with small discs cut from leaves. They divided the leaf disc / soil mixture equally into four test tubes, as shown below.

Leaf disc cut from leaf

Muslin cloth

Leaf discs and soil mixture

A	B	C	D
No added water	No added water	+10ml water	+10ml water
18°C	25°C	18°C	25°C

(a) In which tube would you expect the leaf discs to decay fastest? Give a reason for your answer. .... (2)

....

(b) The students did not add any microorganisms to the test tubes. Where will the microorganisms that cause decay come from? (1)

....

(c) Why did the students seal the tubes with muslin cloth instead of a rubber bung? (1)

....

(d) Suggest **one** way in which the students could use the leaf discs to measure the rate of decay. (1)

....

Recycling

Module 37

For more help on this topic, see Letts GCSE Combined Science Revision Guide page 88

1 Some scientists studied the numbers of cod caught in cool to temperate waters in the northern hemisphere. They obtained the following data, which is expressed in a graph.

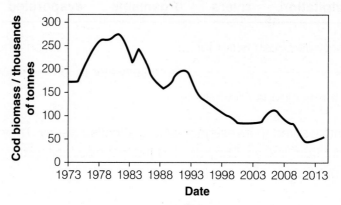

(a) What was the estimated cod biomass in 1988? (1)

.... thousand tonnes

(b) Describe the change in cod numbers between 1988 and 2013. (2)

....

....

(c) International fishing quotas are set in order to manage the numbers of fish in our seas. The table shows data about fishing quotas set by an international fishing commission.

Fish species	UK quota 2013 / tonnes	UK quota 2014 / tonnes
Cod	11 216	13 123
Haddock	27 507	23 381
Whiting	8426	3287

(i) By how much did the cod quota change between 2013 and 2014? (1)

.... tonnes

(ii) Suggest possible reasons for the decreased quota for haddock. (2)

....

....

(d) Suggest **one** other measure that fisheries councils could take to prevent over-fishing. (1)

....

For more help on this topic, see Letts GCSE Combined Science Revision Guide page 89

1 The chemical sodium chloride is more commonly known as 'salt' (i.e. the food flavouring).

Sodium chloride can be made in the laboratory by reacting sodium with chlorine gas. Salt is very soluble in water.

(a) Give the chemical symbols for the two elements present in sodium chloride. Use the periodic table on page 200 to help you. (2)

....

(b) Write a word equation to show the reaction between sodium and chlorine to produce sodium chloride. (2)

....

....

(c) Is sodium chloride a mixture or a compound? Explain your answer. (2)

....

....

(d) When salt dissolves in water, does it form a mixture or a compound? Explain your answer. (2)

....

....

(e) A student wants to separate salt and water from salty water. Which two methods of separation listed below would be appropriate for the student to use? Tick the relevant box or boxes. (2)

Filtration ☐   Simple distillation ☐

Crystallisation ☐   Chromatography ☐

For more help on this topic, see Letts GCSE Combined Science Higher Revision Guide pages 92–93

Atoms and the periodic table

Module 39

1 One of the early representations of the atom, the 'plum pudding' model, was further developed in light of Rutherford, Geiger and Marsden's scattering experiment. This led to the conclusion that the positive charge of an atom is contained within a small volume known as the nucleus. Niels Bohr improved this model and it is this model that forms the basis of the way in which we represent the structure of atoms today.

The plum pudding model

The plum pudding version of an atom is shown above.

Today, an atom such as sodium is represented by a diagram like the one above.

(a) Give **two** differences between the 'plum pudding' model of an atom and today's model. (2)

....

....

(b) What observation in Rutherford, Geiger and Marsden's scattering experiment led them to conclude that the positive charge of an atom was contained in a small volume? (1)

....

(c) What improvements did Niels Bohr make to the nuclear model and what evidence did he have to support these changes? (2)

....

....

(d) James Chadwick developed the idea that the nucleus of an atom contains protons and neutrons.

Complete the table below showing the properties of the sub-atomic particles. (4)

Particle	Relative charge	Relative mass
Proton	+ 1	
Neutron		1
Electron		

(e) An atom of sodium contains 11 protons, 12 neutrons and 11 electrons.

(i) What is the atomic number of sodium? .... (1)

(ii) What is the mass number of sodium? .... (1)

For more help on this topic, see Letts GCSE Combined Science Higher Revision Guide pages 94–95

1 The diagrams below represent an atom of magnesium and an atom of fluorine:

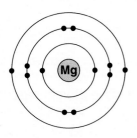 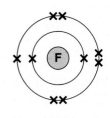

The electronic structure of magnesium can also be written as 2,8,2

(a) How many shells of electrons does a magnesium atom have? (1)

....

(b) What is the electronic structure of fluorine? (1)

....

(c) In which group of the periodic table would you expect to find fluorine?
Explain your answer. (2)

....

....

2 Dmitri Mendeleev is often referred to as the father of the periodic table, as he was
instrumental in its construction.

Mendeleev placed the metals lithium and sodium in the same group of the periodic table.

(a) How are the elements arranged in the periodic table? (1)

....

(b) Explain why Mendeleev placed sodium and lithium in the same group of the
periodic table. (1)

....

(c) Give **two** characteristic properties of metals. (2)

....

....

For more help on this topic, see Letts GCSE Combined Science Higher Revision Guide pages 96–97

Electronic structure and the periodic table

Module 41

1 (a) Explain why the noble gases are chemically inert. (2)

....

....

(b) What is the trend in boiling point as the relative atomic mass of the noble gases

increases? .... (1)

2 The element sodium reacts vigorously with water. When universal indicator solution is added to the resulting solution a colour change is observed.

(a) Write a balanced symbol equation for the reaction between sodium and water. (2)

....

(b) What colour would universal indicator solution turn when it is added to the resulting solution? Explain your answer. (2)

....

....

(c) The element potassium reacts more vigorously with water than sodium. Explain why. (2)

....

....

3 Sodium reacts with bromine gas as shown by the equation below:

........$Na_{(s)}$ + ...$Cl_{2(g)}$ $\longrightarrow$ ....$NaCl_{(s)}$

(a) Balance the above equation. (2)

(b) What type of bonding is present in sodium chloride (NaCl)? (1)

....

(c) When chlorine gas is bubbled through sodium bromide solution a chemical reaction occurs. Write a word equation for this reaction. (2)

....

(d) What name is given to the type of reaction that occurs when chlorine gas reacts with sodium bromide solution? Explain why the reaction occurs. (2)

....

....

For more help on this topic, see Letts GCSE Combined Science Higher Revision Guide pages 98–99

1. Consider the structures of sodium (Na), chlorine (Cl_2) and sodium chloride (NaCl).

Draw arrows from each substance to its correct structure. One structure will not have an arrow drawn to it. (3)

Sodium

Chlorine

Sodium chloride

Ionic

Simple molecular

Giant molecular

Metallic

2. Calcium reacts with oxygen to form calcium oxide. The bonding in calcium oxide is ionic.

(a) Complete the diagrams below to show the electronic configurations and charges of the calcium and oxygen ions. (2)

(b) State the charges on each ion. (2)

....

....

3. (a) Draw a dot-and-cross diagram to show the structure of HCl. (1)

(b) The dot-and-cross diagram for a molecule of oxygen is shown below.

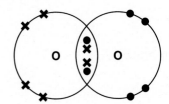

How many covalent bonds are there between the two oxygen atoms? (1)

....

For more help on this topic, see Letts GCSE Combined Science Higher Revision Guide pages 102–103

1 The structure of calcium oxide is shown below.

- ⊖ Negatively charged oxide ions
- ⊕ Positively charged calcium ions

(a) What force holds the ions in place in calcium oxide? (1)

....

(b) What is the empirical formula of calcium oxide? Explain your answer. (2)

....

....

2 The structure of a covalent compound is shown below.

(a) Name this type of structure. .... (1)

(b) What is the empirical formula of this structure? Explain your answer. (2)

....

....

3 The structure of boron nitride is shown below.

- ◯ Boron atoms
- ● Nitrogen atoms

What type of structure does boron nitride have? Explain your answer. (2)

....

....

For more help on this topic, see Letts GCSE Combined Science Higher Revision Guide pages 104–105

1 Solids, liquids and gases are the three main states of matter. The three physical states of matter can be easily interconverted.

(a) Draw a diagram to show the arrangement of particles in a solid. (2)

(b) What forces are broken when a solid turns into a liquid? (1)

....

2 Consider the table below, which provides information about the physical properties of two materials.

	Material A	Material B
Melting point/°C	801	3730
Electrical conductivity of solid	Poor	Poor
Electrical conductivity of liquid	Good	Poor

(a) Material A has a giant ionic structure. Explain how this can be deduced from the table. (1)

....

(b) Explain why material A is a poor conductor of electricity in the solid state. (2)

....

....

(c) Deduce the structure of material B. Explain your answer. (2)

....

....

(d) Explain why material B is a poor conductor of electricity in both the solid and liquid states. (2)

....

....

For more help on this topic, see Letts GCSE Combined Science Higher Revision Guide pages 106–107

1 The diagram shows the structure of a metal such as magnesium.

With reference to the diagram, explain the following:

Free electron

Positive metal ion

(a) Why metals are malleable (i.e. able to be bent and shaped). (2)

....

....

(b) Why metals usually have high melting points. (2)

....

....

(c) Why metals are good conductors of electricity. (2)

....

....

2 Most metals in everyday use are alloys, such as steel.

(a) What is an alloy? (1)

....

(b) With reference to its structure, explain why most alloys are harder than pure metals. (2)

....

....

3 This question is about different structures of carbon.

(a) Give **two** uses of fullerenes. (2)

....

....

(b) State **two** properties of fullerenes. (2)

....

....

(c) What is the difference between a fullerene and a carbon nanotube? (1)

....

For more help on this topic, see Letts GCSE Combined Science Higher Revision Guide pages 108–109

1 Consider the equation below.

$$2Al_{(s)} + Fe_2O_{3(l)} \longrightarrow Al_2O_{3(s)} + 2Fe_{(l)}$$

(a) With reference to the above equation, what do you understand by the term 'conservation of mass'? (1)

....

(b) Complete the table below by filling in the relative atomic (A_r) or formula (M_r) mass of each substance in the equation. (4)

Substance	A_r / M_r
Al	
Fe_2O_3	
Al_2O_3	
Fe	

2 Magnesium oxide can be made by heating magnesium in air or by the thermal decomposition of magnesium carbonate.

(a) Explain why the mass of magnesium increases when it is burnt in air. (2)

....

....

(b) Write a balanced symbol equation for the reaction that occurs when magnesium is heated in air. (2)

....

(c) Explain why the mass of magnesium carbonate decreases when it is heated in air. (2)

....

....

(d) Write a balanced symbol equation for the thermal decomposition of magnesium carbonate. (2)

....

Mass and equations

Module 47

For more help on this topic, see Letts GCSE Combined Science Higher Revision Guide pages 112–113

HT **1** Consider the equation below.

$$4Na_{(s)} + O_{2(g)} \longrightarrow 2Na_2O_{(s)}$$

(a) How many atoms are present in 4 moles of sodium? (1)

....

(b) What is the mass of 4 moles of Na? (1)

....

(c) How many moles are present in 11.5 g of sodium? (1)

....

(d) What mass of sodium oxide is formed when 11.5 g of sodium reacts with oxygen? (2)

....

....

2 A hydrocarbon contains 6 g of carbon and 1 g of hydrogen.

(a) Calculate the empirical formula of this hydrocarbon. (3)

(b) The relative formula mass of this hydrocarbon is 98.

Determine its molecular formula. (2)

....

....

For more help on this topic, see Letts GCSE Combined Science Higher Revision Guide pages 114–115

HT **1** In an experiment, 5 g of sodium hydroxide was dissolved in 200 cm³ of water.

(a) Calculate the concentration of the sodium hydroxide solution in g/dm³. (1)

....

....

(b) What mass of sodium hydroxide is present in 14 cm³ of this solution? (1)

....

....

HT **2** Lead forms the following oxides: PbO, PbO_2 and Pb_3O_4. In an experiment, 41.4 g of lead reacts with 3.2 g of oxygen to form 44.6 g of PbO.

Complete the table below to work out the equation for the reaction of lead with oxygen to form lead oxide. (4)

Chemical	Pb	O_2	PbO
Mass from question/g	41.4	3.2	44.6
A_r or M_r	207		223
Moles = $\dfrac{mass}{M_r}$	$\dfrac{41.4}{207}$ $= 0.2$		
÷ smallest			

Balanced equation: .... $Pb + O_2 \longrightarrow$ .... PbO

For more help on this topic, see Letts GCSE Combined Science Higher Revision Guide pages 116–117

1 Calcium reacts with oxygen present in the air to form calcium oxide.

(a) Write a word equation for this reaction. (2)

....

(b) Explain why this is an oxidation reaction. (2)

....

....

(c) Name a metal that could be used to displace calcium from calcium oxide.
Explain your choice of metal. (2)

....

....

(d) Write a word equation for this reaction. (2)

....

(e) In your equation from part **d**, give the name of the substance that has been reduced. (1)

....

2 Many metals are found in the Earth's crust as ores. For example, haematite is an ore containing iron(III) oxide (Fe_2O_3); bauxite is an ore containing aluminium oxide (Al_2O_3). Both metals can be extracted from their ores by reduction.

(a) What process is used to extract metals above carbon in the reactivity series from

their ores? .... (1)

(b) When iron(III) oxide is reduced by carbon, iron and carbon dioxide are formed.
Write a balanced symbol equation for this reaction. (2)

....

HT **(c)** In the extraction of aluminium the following equation occurs: $Al^{3+} + 3e^- \longrightarrow Al$

Is this an oxidation or reduction reaction? Explain your answer. (3)

....

....

HT **(d)** Aluminium can also be formed by reacting a reactive metal such as potassium.
An ionic equation for this reaction would be: $3K + Al^{3+} \longrightarrow Al + 3K^+$
Which substance is oxidised in this reaction? Explain your answer. (2)

....

For more help on this topic, see Letts GCSE Combined Science Higher Revision Guide pages 120–121

1 Zinc, zinc oxide and zinc carbonate all react with acids.

(a) Name the salt formed when zinc, zinc oxide and zinc carbonate react with sulfuric acid.

(1)

....

(b) Write a word equation for the reaction between zinc oxide and hydrochloric acid. (2)

....

(c) Name and give the formula of the gas formed when zinc carbonate reacts with nitric acid. (2)

....

HT **(d)** The ionic equation for the reaction between calcium and acid is:

$$Ca_{(s)} + 2H^+_{(aq)} \longrightarrow Ca^{2+}_{(aq)} + H_{2(g)}$$

(i) Which substance is oxidised in the above reaction? (1)

....

(ii) Which substance is reduced? (1)

....

HT **(e)** In terms of electrons, what is meant by an oxidation reaction? (2)

....

2 Copper(II) oxide (CuO) reacts with sulfuric acid to make the soluble salt copper(II) sulfate according to the equation below.

$$CuO_{(s)} + H_2SO_{4(aq)} \longrightarrow CuSO_{4(aq)} + H_2O_{(l)}$$

(a) Describe the steps you would take to prepare a sample of solid copper(II) sulfate, starting from copper(II) oxide and sulfuric acid solution. (3)

....

....

....

(b) The salt sodium nitrate can be formed by the reaction of sodium carbonate with an acid. Name this acid. (1)

....

(c) Is sodium nitrate a soluble or insoluble salt? .... (1)

For more help on this topic, see Letts GCSE Combined Science Higher Revision Guide pages 122–123

Reactions of acids

Module 51

1 Hydrogen chloride (HCl) and ethanoic acid (CH_3COOH) both dissolve in water to form acidic solutions.

(a) What is the pH range of acids? (1)

....

(b) Give the name and formula for the ion present in solutions of hydrogen chloride and ethanoic acid. (2)

....

(c) What is the common name for hydrogen chloride solution? (1)

....

(d) Acids can be neutralised by reaction with an alkali such as sodium hydroxide.

Write an ionic equation for the reaction that occurs when an acid is neutralised by an alkali. Include state symbols in your equation. (2)

....

HT **2** Ethanoic acid is a key ingredient in vinegar. It is a weak acid. Nitric acid is a strong acid.

(a) What is the difference between a weak acid and a strong acid? (2)

....

....

(b) Write a balanced symbol equation for the dissociation of ethanoic acid in water. (2)

....

(c) Will a solution of nitric acid of the same concentration as a solution of ethanoic acid have a higher or lower pH?

Explain your answer. (2)

....

....

(d) What happens to the pH value of an acid when it is diluted by a factor of 10? (2)

....

....

For more help on this topic, see Letts GCSE Combined Science Higher Revision Guide pages 124–125

1 The diagram shows how a molten salt such as lead(II) bromide ($PbBr_2$) can be electrolysed. Lead(II) bromide consists of Pb^{2+} and Br^- ions.

Molten lead(II) bromide

(a) What general name is given to positive ions? (1)

....

(b) What general name is given to negative ions?

.... (1)

(c) Explain why the lead(II) bromide needs to be molten. (2)

....

....

HT **(d)** At the anode, the following reaction occurs: $2Br^- \longrightarrow Br_2 + 2e^-$

Is this an oxidation or a reduction reaction? Explain your answer. (2)

....

....

2 **(a)** The table below shows the products at each electrode when the following solutions are electrolysed. Some answers have already been filled in. Complete the table. (6)

Solution	Product at anode	Product at cathode
NaCl	H_2	
KNO$_3$		
CuSO$_4$		
Water diluted with sulfuric acid		O_2

(b) Explain why sodium is not formed at the cathode when aqueous sodium chloride is electrolysed. (2)

....

....

HT **(c)** Write a half-equation for the formation of oxygen at the anode in the electrolysis of water diluted with sulfuric acid. (2)

....

Applications of electrolysis

Module 53

For more help on this topic, see Letts GCSE Combined Science Higher Revision Guide pages 126–127

1 When methane burns in air an exothermic reaction takes place. The equation for the reaction is shown here:

$$CH_{4(g)} + 2O_{2(g)} \longrightarrow CO_{2(g)} + 2H_2O_{(l)}$$

(a) On the axes, draw and label a reaction profile for an exothermic reaction. Label the reactants, products, activation energy and ΔH. (5)

(b) In terms of bond energies, explain why this reaction is exothermic. (2)

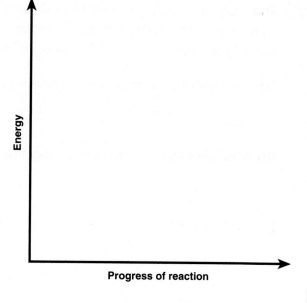

....

....

(c) Give an example of an endothermic chemical reaction. (1)

....

HT **2** Hydrogen peroxide (H_2O_2) decomposes in air to form water and oxygen. The diagram below shows this reaction by displaying all of the bonds present in the reactants and products.

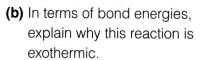

The table states the bond energy values of the bonds present in hydrogen peroxide, water and oxygen.

Bond	Bond energy kJ/mol
H–O	463
O–O	146
O=O	496

For the decomposition of hydrogen peroxide:

(a) Calculate ΔH for the reaction. (3)

....

....

(b) Is this reaction exothermic or endothermic? Explain your answer. (2)

....

....

For more help on this topic, see Letts GCSE Combined Science Higher Revision Guide pages 128–129

1 A pupil was investigating the effect of temperature on the rate of reaction between aqueous sodium thiosulfate with dilute hydrochloric acid. When the acid is added to the sodium thiosulfate solution, a precipitate of sulfur gradually forms. The pupil recorded the time taken for a cross written on a piece of paper to disappear from view.

The experiment was repeated at different temperatures. The results are shown in the table below.

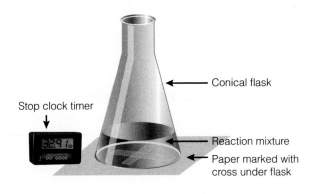

Temperature (°C)	Time taken for cross to disappear from view (s)
15	70
25	44
40	30
55	22
70	14

(a) At which temperature was the reaction the fastest? .... (1)

(b) Suggest how the rate of this reaction at 30°C will change when the concentration of hydrochloric acid is increased. Explain your answer. (2)

....

....

2 A student investigating the rate of reaction between magnesium and hydrochloric acid carried out two experiments: one using magnesium ribbon and the other using magnesium powder.

The equation for the reaction is shown here: $Mg_{(s)} + 2HCl_{(aq)} \longrightarrow MgCl_{2(aq)} + H_{2(g)}$

In the first experiment, 69 cm³ of hydrogen gas was collected in 46 seconds. In the second experiment, 18 cm³ of hydrogen gas was collected in 10 seconds.

(a) Calculate the rate of reaction in both experiments. (3)

....

....

(b) Which experiment was carried out using magnesium powder? Explain your answer. (2)

....

Rates of reaction

Module 55

For more help on this topic, see Letts GCSE Combined Science Higher Revision Guide pages 132–133

1 One aspect of collision theory states that, for a chemical reaction to occur, the reacting particles must first collide with each other.

(a) What else must occur in order for the collision to be successful? (1)

....

(b) Explain how:

    (i) Increasing the pressure of a gas increases the rate of a reaction. (2)

....

....

    (ii) Increasing the temperature of a solution increases the rate of a reaction. (3)

....

....

....

(c) Adding a catalyst to a reaction increases the rate of reaction. Draw, on the diagram below, the reaction profile for the reaction when a catalyst is added. (1)

(d) Explain how a catalyst is able to increase the rate of a reaction. (2)

....

....

For more help on this topic, see Letts GCSE Combined Science Higher Revision Guide pages 134–135

1 In an experiment, a student heated some blue crystals of hydrated copper(II) sulfate in an evaporating dish.

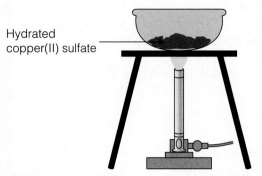

Hydrated copper(II) sulfate

(a) State what colour change will be observed during the reaction. (1)

....

(b) Write a word equation for the change that occurs to the hydrated copper(II) sulfate during this experiment. (2)

....

(c) Describe what will be observed when water is added to the solid remaining at the end of the experiment. (1)

....

(d) Is the addition of water an exothermic or endothermic reaction? (1)

....

2 Consider the reaction given here: $2SO_{2(g)} + O_{2(g)} \rightleftharpoons 2SO_{3(g)}$

(a) What does the $\rightleftharpoons$ symbol represent? .... (1)

HT **(b)** What would be the effect on the yield of SO_3 if the above reaction was carried out at a higher pressure? Explain your answer. (3)

....

....

....

HT **(c)** The forward reaction is exothermic. What would be the effect on the yield of SO_3 if the above reaction was carried out at a higher temperature? Explain your answer. (3)

....

....

....

For more help on this topic, see Letts GCSE Combined Science Higher Revision Guide pages 136–137

1 Crude oil is separated into its constituent components by fractional distillation. A fractionating column is used to carry out fractional distillation.

A diagram of a fractionating column is shown here.

Refinery gases

Gasoline (petrol)

X

Diesel oil

Fuel oil

Heated crude oil →

Bitumen

(a) What is the name of the fraction labelled as 'X'?   (1)

....

(b) As you go down the fractionating column, what happens to the boiling point of the fractions?   (1)

....

(c) Explain how crude oil is separated in a fractionating column.   (3)

....

....

....

2 Most of the hydrocarbons found in crude oil are members of the homologous series of hydrocarbons called the alkanes.

(a) What is the general formula of alkanes?   (1)

....

(b) In the space below, draw the displayed formula for propane.   (1)

(c) What is the molecular formula of the member of the homologous series after propane?   (1)

....

(d) Other than the same general formula, give one other feature of a homologous series.   (1)

....

For more help on this topic, see Letts GCSE Combined Science Higher Revision Guide pages 140–141

1 Propane (C_3H_8) is an alkane frequently bottled and used as a fuel.

(a) Name the products produced when propane undergoes complete combustion. (2)

....

....

(b) Write a balanced symbol equation for the complete combustion of propane. (2)

....

....

(c) In the complete combustion of propane which substance has been oxidised? (1)

....

2 Cracking is a method of converting long-chain hydrocarbons into shorter, more useful hydrocarbons.

(a) Name a catalyst used to crack hydrocarbons. (1)

....

(b) Other than use of a catalyst, state an alternative method for cracking hydrocarbons. (2)

....

....

(c) The equation for a reaction that occurs during this process is:

$$C_{12}H_{26} \longrightarrow C_2H_4 + C_6H_{12} + X$$

In the balanced equation, what is the molecular formula of X? (1)

....

(d) Describe how the gas collected during cracking can be shown to contain a double bond. (2)

....

....

For more help on this topic, see Letts GCSE Combined Science Higher Revision Guide pages 142–143

Combustion and cracking of hydrocarbons

Module 59

1 Paint and household cleaning chemicals are typical examples of everyday chemical formulations.

(a) What is a formulation? (1)

....

(b) Are formulations chemically pure? Explain your answer. (2)

....

....

(c) Give **one** other example of a type of formulation. (1)

....

2 Pen ink is typically a mixture of dyes. The individual dyes in pen ink can be separated by paper chromatography. The chromatogram below is for an unknown ink 'X' and the standard colours blue, red and green.

(a) What is the mobile phase in paper chromatography? (1)

....

(b) Which colours are present in ink X? Explain your answer. (2)

....

....

(c) Calculate the R_f for the blue ink. (2)

....

....

(d) The sample of ink could also have been separated using gas chromatography. State **one** advantage that gas chromatography has over paper chromatography. (1)

....

For more help on this topic, see Letts GCSE Combined Science Higher Revision Guide pages 146–147

1 When carbon dioxide gas is bubbled through limewater the following reaction occurs:

$$Ca(OH)_{2(aq)} + CO_{2(g)} \longrightarrow CaCO_{3(s)} + H_2O_{(l)}$$

(a) What is the chemical name for limewater? (1)

....

(b) With reference to the above equation, explain why limewater turns cloudy when carbon dioxide is bubbled through it. (2)

....

....

2 (a) Describe what would be observed when moist blue litmus paper is added to a gas jar containing chlorine gas. (2)

....

....

(b) Is chlorine an acidic or alkaline gas?

What evidence is there to support your answer? (2)

....

....

3 Hydrogen and oxygen gases are both colourless, odourless gases.

(a) What is the chemical test and observation for hydrogen gas? (2)

....

....

(b) What is the chemical test and observation for oxygen gas? (2)

....

....

(c) When hydrogen gas burns in air it reacts with oxygen to form water. Write a balanced symbol equation for the reaction taking place. (2)

....

....

Identification of gases

Module 61

For more help on this topic, see Letts GCSE Combined Science Higher Revision Guide pages 148–149

1 Many scientists think that the Earth's early atmosphere may have been similar in composition to the gases typically released by volcanoes today.

The pie charts below show the composition of the atmosphere today and the composition of gases released by a volcano.

Composition of Earth's atmosphere today

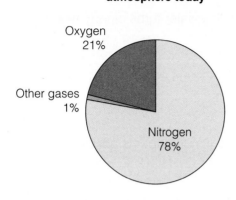

Oxygen 21%

Other gases 1%

Nitrogen 78%

Composition of gases from volcanoes

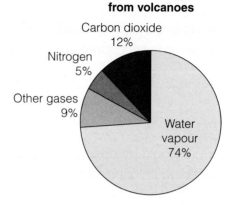

Carbon dioxide 12%

Nitrogen 5%

Other gases 9%

Water vapour 74%

(a) Name a gas that is more abundant in the Earth's atmosphere today than in the early atmosphere. (1)

....

(b) Why is the amount of water vapour in the atmosphere significantly less now than in the early atmosphere? (2)

....

....

(c) Describe the process that caused the percentage of oxygen in the atmosphere to increase. (2)

....

....

(d) Outline **two** ways in which the amount of carbon dioxide in the early atmosphere decreased. (2)

....

....

For more help on this topic, see Letts GCSE Combined Science Higher Revision Guide pages 152–153

1 When fossil fuels are burnt, carbon dioxide gas is produced and released into the atmosphere. Some people believe carbon dioxide to be a greenhouse gas.

(a) Explain how greenhouse gases maintain the temperature on Earth. (2)

....

....

(b) Name **one** other greenhouse gas. (1)

....

(c) Outline **two** ways in which human activity increases the amounts of greenhouse gases in the atmosphere. (2)

....

....

(d) Why is it not easy to predict the impact of changes on global climate change? (2)

....

....

(e) Describe **two** potential effects of increasing average global temperature. (2)

....

....

(f) Describe **two** forms of global action that can be taken to reduce the carbon footprint. (2)

....

....

(g) Outline **two** potential problems that governments might face when trying to reduce the carbon footprint. (2)

....

....

For more help on this topic, see Letts GCSE Combined Science Higher Revision Guide pages 154–155

1 The combustion of fossil fuels, e.g. from cars, is a major source of atmospheric pollution.

(a) What is meant by a fossil fuel? (1)

....

(b) How are carbon and carbon monoxide formed from the combustion of fossil fuels? (1)

....

(c) State **one** harmful effect caused by the release of oxides of nitrogen into the atmosphere. (1)

....

(d) Give two detrimental consequences of releasing particulates into the air. (2)

....

....

(e) Explain why carbon monoxide is a toxic gas. (2)

....

....

(f) Fuels can contain sulfur as an impurity. Explain how burning this impurity can cause problems in the environment. (3)

....

....

....

For more help on this topic, see Letts GCSE Combined Science Higher Revision Guide pages 156–157

1. We use the Earth's resources to provide us with warmth, shelter, food and transport. As the Earth is a finite source of resources, we need to ensure that we use them sustainably.

   (a) What is sustainable development? (1)

   ....

   (b) What is meant by the term 'potable water'? (1)

   ....

   (c) After identification of an appropriate source of fresh water, what two stages then need to occur to turn this into potable water? (2)

   ....

   ....

   (d) Outline the processes that are used to treat sewage. (3)

   ....

   ....

   ....

2. Salty water, e.g. seawater, can be desalinated (i.e. treated to reduce its salt content). One such method is distillation.

   (a) In the space below, draw a labelled diagram to show how seawater can be distilled. Describe how distillation produces desalinated water. (5)

   ....

   ....

   (b) What method of desalination involves the use of membranes? (1)

   ....

For more help on this topic, see Letts GCSE Combined Science Higher Revision Guide pages 158–159

HT ❶ As we mine more and more of the Earth's natural resources, we need to develop alternative methods of extracting metals such as copper. One such recently developed method is phytomining.

(a) What is an ore? (2)

....

....

(b) Aside from environmental reasons, why are methods of extracting metals such as phytomining being developed? (2)

....

....

(c) State two uses of copper and describe the properties that make it suitable for each purpose. (4)

....

....

(d) Describe the process of phytomining. (3)

....

....

....

(e) How can bacteria be used to extract metals such as copper? (1)

....

(f) Scrap metals, such as iron, can be used to obtain copper from solutions. Write a word equation for the reaction that occurs when iron reacts with copper sulfate solution. (2)

....

(g) Why would platinum not be a suitable metal for extracting copper from copper sulfate solution? (1)

....

(h) What is the name of the process involving electricity that can be used to extract a metal from a solution of its ions? (1)

....

For more help on this topic, see Letts GCSE Combined Science Higher Revision Guide pages 160–161

1 Life-cycle assessments (LCAs) are carried out to evaluate the environmental impact of products at different stages of the life of a product.

(a) Other than transport and distribution, state **two** other stages of a product life cycle that will be considered when carrying out an LCA. (2)

....

....

(b) State **two** quantities that are considered when conducting an LCA. (2)

....

....

(c) Explain why the pollution value in a life-cycle value may result in the LCA not being totally objective. (2)

....

....

(d) Why might selective or abbreviated LCAs be misused? (1)

....

2 The table below shows an example of an LCA for the use of plastic (polythene) and paper shopping bags.

	Amounts per 1000 bags over the whole LCA	
	Paper	**Plastic (polythene)**
Energy use (mJ)	2590	713
Fossil fuel use (kg)	28	12.8
Solid waste (kg)	34	6
Greenhouse gas emissions (kg CO_2)	72	36
Fresh water use (litres)	3387	198

Based on the above figures, why might some people think that plastic (polythene) bags should be used instead of paper bags? Use information from the table in your answer. (3)

....

....

....

For more help on this topic, see Letts GCSE Combined Science Higher Revision Guide pages 162–163

Module 67

1 Neil, an astronaut, has a mass of 78 kg.

(a) Calculate Neil's weight.

(gravitational field strength = 10 N/kg) (1)

....

(b) The Moon has a gravitational field strength of 1.6 N/kg.

What would Neil's mass and weight be on the Moon? (2)

....

....

(c) Neil also has a gravitational field that causes attraction.

Explain why this has no measurable effect on either the Moon or the Earth. (2)

....

....

2 The diagram shows Louise, a runner.

(a) Calculate the resultant force acting on Louise.

30 N   120 N

(2)

....

....

(b) The forces on Louise changed as shown by the diagram below.

(i) What is the resultant force on Louise now?

90 N   90 N

(1)

....

(ii) What statement can be made about her speed now? Explain your answer. (2)

....

....

(c) Louise uses 90 N of force and the work done is 25 kJ. What distance does she cover? (2)

....

....

For more help on this topic, see Letts GCSE Combined Science Higher Revision Guide pages 166–167

1 The graph below shows the results of an extension of a spring in response to different forces.

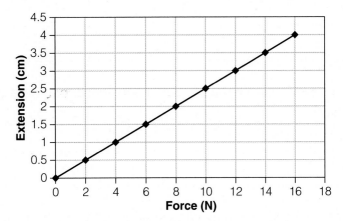

(a) What term is given to the relationship shown in this graph? (1)

....

(b) Calculate the spring constant for this spring. (3)

....

....

....

(c) The results were repeated, but with a larger range of masses providing the force. The results are shown in the graph below.

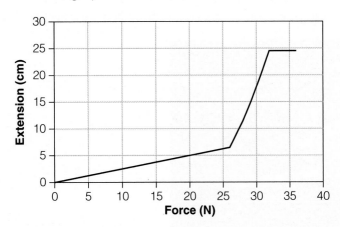

(i) What term is given to the relationship shown in this graph? (1)

....

(ii) Explain the difference in the shapes of the two graphs. (3)

....

....

....

For more help on this topic, see Letts GCSE Combined Science Higher Revision Guide pages 168–169

1 (a) Explain why speed is a scalar quantity while velocity is a vector quantity. (2)

....

....

HT (b) Why does a car travelling around a bend at a constant speed have a constantly changing velocity? (2)

....

....

(c) Draw lines to match the following average speeds to the methods of travel. (4)

Average speed	Travel method
1.5 m/s	Car
6 m/s	Plane
20 m/s	Bicycle
250 m/s	Walking

2 (a) A jet ski travels 180 metres in 13 seconds. What is its speed? (2)

....

....

(b) How long would it take for the same jet ski travelling at the same speed to travel 500 m? (2)

....

....

For more help on this topic, see Letts GCSE Combined Science Higher Revision Guide pages 170–171

1 The graph shows the distance a boat travels over time.

(a) What is the average speed of the boat over the first 350 seconds? (2)

....

....

....

....

(b) What is the velocity of the boat at 265 seconds? (2)

....

....

2 The graph shows the velocity of a car.

(a) (i) During what time period is the greatest acceleration of the car? (1)

....

....

(ii) What is the acceleration during this time period? (2)

....

....

(HT) (b) How far did the car travel in 0–50 seconds? (3)

....

....

3 Explain why a skydiver in free fall accelerates for a time before maintaining a

constant speed. .... (4)

....

....

....

For more help on this topic, see Letts GCSE Combined Science Higher Revision Guide pages 172–173

1 **(a)** What will happen if a resultant force greater than zero is applied to:

(i) a stationary object? (1)

....

(ii) a moving object? (1)

....

(b) Which of Newton's laws deals with the effect of resultant force? (1)

....

HT **(c)** What term is given to the tendency of objects to continue in their state of rest? (1)

....

2 **(a)** The acceleration of an object is inversely proportional to the mass of an object.
How does the acceleration relate to the resultant force? (1)

....

(b) (i) Adele, a cyclist, who has a total mass of 110 kg, accelerates at 5 m/s^2.
What is the resultant force on Adele? (2)

....

....

(ii) Explain how the resultant force on Adele will change as she slows down and
comes to a stop. (3)

....

....

....

3 State Newton's third law. (1)

....

4 Use Newton's laws to explain the events that the sign below is warning about. (3)

....

....

....

....

**Weak bridge
¼ mile
ahead**

For more help on this topic, see Letts GCSE Combined Science Higher Revision Guide pages 174–175

1 (a) Increasing speed increases the thinking distance and the braking distance of a car. Explain why. (2)

(i) Thinking distance: ....

....

(ii) Braking distance: ....

....

(b) Give **two** other factors that could increase the thinking distance. (2)

....

....

(c) The table shows the temperatures of brakes when a car decelerates to zero in a set amount of time. The test was repeated at four different initial speeds.

Test	Brake temperature (°C)
A	150
B	800
C	320
D	560

(i) Put the tests in order of initial speed, from highest speed to lowest speed. (1)

....

(ii) Explain the difference in the results of A and B. Use the concepts of work done and kinetic energy in your answer. (3)

....

....

....

HT 2 (a) What is the momentum of a 50 kg cheetah running at 20 m/s? (2)

....

....

(b) What will happen to the cheetah's momentum as it decelerates? (1)

....

Forces, braking and momentum

Module 73

For more help on this topic, see Letts GCSE Combined Science Higher Revision Guide pages 176–177

1. Jessica, a sky diver, is in a plane at a height of 3900 m. She has a mass of 60 kg.

   (a) What is Jessica's gravitational potential energy?

   (Assume gravitational field strength is 10 N/kg.) (2)

   ....

   ....

   (b) Jessica jumps from the plane and accelerates to a terminal velocity of 55 m/s.
   What is Jessica's kinetic energy at this point? (2)

   ....

   ....

   (c) Jessica deploys her parachute and slows to a speed of 5 m/s.
   What is her new kinetic energy? (2)

   ....

   ....

   (d) When Jessica lands on the ground, she is stationary. What is her kinetic energy now? (1)

   ....

2. The table below shows the specific heat capacity of different substances.

Substance	Specific heat capacity J/kg/°C
Water	4180
Copper	390
Ethanol	2440
Titanium	520

   (a) Which substance would require the least energy to raise 1 kg by 1°C?
   Explain how you arrived at your answer. (2)

   ....

   ....

   (b) Calculate the mass of copper that requires 76 kJ of energy to increase in temperature
   from 16°C to 35°C. (3)

   ....

   ....

   ....

For more help on this topic, see Letts GCSE Combined Science Higher Revision Guide pages 180–181

1 Which of the below is an incorrect ending to the following sentence?
Tick the correct options. (2)

Energy cannot be...

created. ☐

transferred. ☐

destroyed. ☐

dissipated. ☐

2 The table below shows the average cost of heating three different houses.
Each home is the same size.

House	Heating cost per hour / p
A	105
B	88
C	150

(a) Which house is likely to have the best insulation in its walls?
Explain how you arrived at your answer. (3)

....

....

....

(b) Can you be certain about your conclusion?
Explain your answer. (3)

....

....

....

(c) The boiler in another house is 65% efficient. If the energy input is 5400 kJ,
what is the useful energy output? (2)

....

....

For more help on this topic, see Letts GCSE Combined Science Higher Revision Guide pages 182–183

1 Complete the table below to give **three** examples of renewable energy and **three** examples of non-renewable energy. (3)

Renewable energy	Non-renewable energy

2 The graph below shows the percentage of the total electricity in the UK generated by wind turbines.

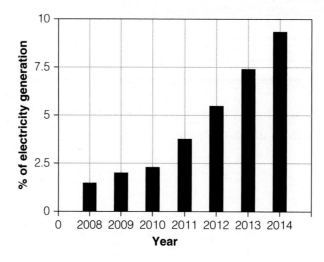

(a) Describe the trend shown by the graph. (1)

....

(b) Describe the main advantage of wind power over burning fossil fuels. (2)

....

....

(c) Will wind turbines ever make up 100% of electricity generation? Explain your answer. (3)

....

....

....

(d) Why might some people not want wind turbines to be placed on the tops of hills? (1)

....

For more help on this topic, see Letts GCSE Combined Science Higher Revision Guide pages 184–185

1 The speed of sound in water is 1482 m/s.

(a) What type of wave is a sound wave when travelling in air? (1)

....

(b) Calculate the wavelength of a sound wave in water that has a frequency of 120 Hz. (2)

....

....

(c) The speed of sound in air is 330 m/s.
Suggest a reason for the difference between this speed and the speed of sound
in water. (2)

....

....

(d) How do the oscillations in a water wave differ to the oscillations in a sound wave
travelling in air? (2)

....

....

2 Draw lines to match the following terms to their correct definitions. (3)

| Amplitude |

| The distance from a point on one wave to the equivalent point on the adjacent wave. |

| Wavelength |

| The number of waves passing a point each second. |

| Frequency |

| The maximum displacement of a point on a wave away from its undisturbed position. |

For more help on this topic, see Letts GCSE Combined Science Higher Revision Guide pages 188–189

1 The image below shows the electromagnetic spectrum.

Wavelength (metres)	**A**	Microwave	**B**	Visible	Ultraviolet	X-ray	**C**
	10^3	10^{-2}	10^{-5}	10^{-6}	10^{-8}	10^{-10}	10^{-12}

What are the names of the different types of electromagnetic radiation **A–C**? (3)

A: ....

B: ....

C: ....

HT **2** The diagram shows the effect of light entering water from the air.

(a) Give the names of angles **A** and **B**. (2)

A: ....

B: ....

(b) Explain why the light changes direction as it moves from air to water. (3)

....

....

....

(c) How would refraction differ if the light was entering air from water? (1)

....

For more help on this topic, see Letts GCSE Combined Science Higher Revision Guide pages 190–191

HT **1** Draw lines to match up the following types of radiation to their uses and why they are suitable for that use. (3)

Radiation type	Use	Why suitable for use?
Radio waves	Heating a room	Don't require a direct line of sight between transmitter and receiver.
Ultraviolet	Television	Require less energy than conventional lights.
Infrared	Energy efficient lamps	Thermal radiation heats up objects.

2 **(a)** From where do gamma rays originate? (1)

....

(b) Give **one** potential use of gamma rays. (1)

....

(c) Why are gamma rays potentially harmful? (2)

....

....

3 What variable could be measured to determine the possible harm caused by radiation? (1)

....

HT **4** This photograph shows a medical image.

(a) What type of electromagnetic radiation is used to produce this image? (1)

....

(b) Explain why this type of radiation is used in medical imaging. (3)

....

....

....

For more help on this topic, see Letts GCSE Combined Science Higher Revision Guide pages 192–193

1 Identify the following circuit symbols:

(a) ────▭──── (1)

....

(b) ────▭/──── (1)

....

(c) ──o─/─o── (1)

....

2 The diagram shows a circuit.

(a) Identify components **X**, **Y** and **Z**. (3)

X: ....

Y: ....

Z: ....

(b) Ammeter **1** gave a reading of 5 amps.
What is the reading of ammeter **2**? (1)

....

(c) What is 'electric current'? (1)

....

(d) If the charge flow through the circuit was 780 C, for how long was the current
flowing through the circuit? (2)

....

....

For more help on this topic, see Letts GCSE Combined Science Higher Revision Guide pages 196–197

1 The graph shows the relationship between the current and potential difference of a component.

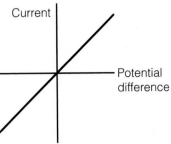

(a) What general name could be given to this type of conductor? (1)

....

(b) What environmental factor could change and lead to a change in the shape of the graph above? (1)

....

(c) On the axis, sketch a graph to show the relationship between current and potential difference through a diode. (2)

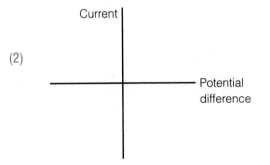

2 Calculate the current flowing through a 4 Ω resistor when there is a potential difference of 15 V across it. (2)

....

....

3 Look at the circuit below.

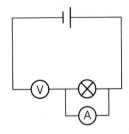

Explain why this circuit could not be used to determine the resistance of the bulb. (3)

....

....

....

For more help on this topic, see Letts GCSE Combined Science Higher Revision Guide pages 198–199

1 The diagram shows a circuit.

(a) What type of circuit is shown? (1)

....

....

4 Ω   5 Ω   2 Ω

(b) What is the total resistance of the three resistors in the circuit? Show how you arrived at your answer. (2)

....

....

2 The diagram shows a circuit.

(a) What is the current at **1**? (1)

....

....

(b) What is the voltage at **2**? (1)

....

....

(c) What is the resistance at **3**? (1)

....

....

(d) The circuit was rewired so that the bulb and resistor were wired in series. How would this change the total resistance of the two components?

Explain your answer. (3)

....

....

....

For more help on this topic, see Letts GCSE Combined Science Higher Revision Guide pages 200–201

1 (a) Draw lines to match the names of the wires in a plug to its colour and function. (3)

Wire	Colour	Function
Live wire	Blue	Completes the circuit.
Neutral wire	Brown	Only carries a current if there is a fault.
Earth wire	Green and yellow stripes	Carries the alternating potential difference from the supply.

(b) Why does a double insulated appliance not require an earth connection? (2)

....

....

2 An electric sander is being used during a house renovation. The sander is plugged into a socket in the wall.

(a) Is the sander powered by direct current or alternating current?
Explain your answer. (2)

....

....

(b) When using the sander, a wire in a wall is damaged and trips the house circuit breakers.

What are the advantages, in this case, of the house mains supply having a circuit breaker rather than a fuse? (2)

....

....

For more help on this topic, see Letts GCSE Combined Science Higher Revision Guide pages 202–203

1 (a) A tumble dryer has a current flowing through it of 3 A and a resistance of 76.7 Ω.

Calculate its power. (2)

....

....

(b) What useful energy transfers are occurring in the tumble dryer? (2)

....

....

(c) If the tumble dryer transfers 180 kJ of energy in a cycle, how long is the cycle in seconds? (2)

....

....

(d) The mains electricity used by the tumble dryer has a potential difference of 230 V. Some of the overhead cables that carry the electricity in the National Grid have a potential difference of 138 kV.

Explain the reasons for this difference in potential difference. (3)

....

....

....

For more help on this topic, see Letts GCSE Combined Science Higher Revision Guide pages 204–205

1 Complete the diagram below to show the poles of the three bar magnets interacting. (3)

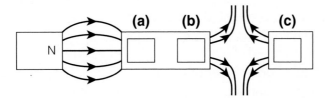

2 The diagram below shows an experiment into magnets and magnetic objects.

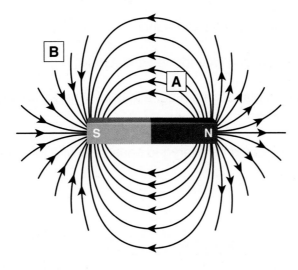

(a) What name is given to the region around the magnet where magnetic objects or magnets experience a force? (1)

....

(b) What type of force would a magnetic object placed at **A** experience? (1)

....

(c) Would the force at **A** be stronger than the force experienced by a magnetic object at **B**? Explain your answer. (2)

....

....

3 The bar magnet in a compass points towards magnetic north. What can be concluded about the Earth's core from this fact? (3)

....

....

....

For more help on this topic, see Letts GCSE Combined Science Higher Revision Guide pages 208–209

1. Tick each of the following statements that correctly describe the features of a solenoid. (2)

   The magnetic field around a solenoid is the same shape as the magnetic field around a bar magnet. ☐

   Adding an iron core to a solenoid reduces the magnetic strength of the solenoid. ☐

   A solenoid with an iron core is an electromagnet. ☐

   A solenoid with a tin core is an electromagnet ☐

   A solenoid is a non-conducting wire formed into a helix. ☐

HT 2. A conducting wire is placed in a magnetic field. When a current is passed through the wire it begins to move.

   (a) What **two** things could be altered to cause the wire to move in the opposite direction? (2)

   ....

   ....

   (b) What name is given to this effect? (1)

   ....

   (c) The wire is at a right angle to the magnetic field. The wire is 30 cm long and has a current of 15 A flowing through. This produces 25 N of force.

   What is the magnetic flux density? (2)

   ....

   ....

HT 3. In an electric fan, electrical energy is transferred to kinetic energy. Explain how this is made possible by the motor effect. (3)

   ....

   ....

   ....

For more help on this topic, see Letts GCSE Combined Science Higher Revision Guide pages 210–211

1 (a) A full fire extinguisher has a mass of 5 kg and a volume of 0.002 m³.
What is the density of a fire extinguisher? (2)

....

....

(b) What would happen to the density of the fire extinguisher after it had been used to
put out a fire? Explain your answer. (3)

....

....

....

2 In an investigation into the link between pressure and molar concentration of a gas,
the temperature and volume are kept constant.

Explain the importance of controlling the following variables:

(a) Constant temperature (1)

....

(b) Constant volume (1)

....

3 (a) What mass of water is produced by heating 3 kg of ice until it has completely melted?
Explain how you arrived at your answer. (2)

....

....

(b) The temperature of the water is then lowered past its freezing point and ice reforms.
How does this prove that the original melting was a physical change as opposed to a
chemical change? (2)

....

....

....

The particle model and pressure

Module 87

For more help on this topic, see Letts GCSE Combined Science Higher Revision Guide pages 214–215

1 The graph below shows the results of an investigation into the heating of a substance.

(a) At what temperatures did the substance do the following?

(i) Melt: .... (1)

(ii) Boil: .... (1)

(b) Explain how you arrived at your answers for part **(a)**. (3)

....

....

....

(c) The substance has a specific latent heat of melting of 250 kJ/kg. What is the energy
required to melt 500 g of the substance? (2)

....

....

2 **(a)** What is 56°C in Kelvin? .... (1)

(b) Explain why it is impossible for an object to have a temperature of −300°C. (2)

....

....

....

For more help on this topic, see Letts GCSE Combined Science Higher Revision Guide pages 216–217

1 An element symbol is shown below.

(a) State the numbers of the following subatomic particles that an atom of the element contains.

    (i) Protons: .... (1)

    (ii) Neutrons: .... (1)

    (iii) Electrons: .... (1)

(b) The element is also found in the following forms:

^{12}**C** ^{13}**C** ^{14}**C**

    (i) What name is given to these forms? (1)

    ....

    (ii) How do they differ from each another? (1)

    ....

2 **(a)** Give **two** differences between the plum pudding model of atomic structure and Rutherford, Geiger and Marsden's nuclear model. (2)

....

....

(b) Why was James Chadwick's experimental work in 1932 important in further developing the model of atomic structure? (1)

....

For more help on this topic, see Letts GCSE Combined Science Higher Revision Guide pages 220–221

1 An investigation was carried out into the penetration of radiation emitted from different radioactive sources.

Type of radiation	Radiation released	Penetration distance in air
A	Two neutrons and two protons (a helium nucleus).	A few centimetres.
B	Electromagnetic radiation from the nucleus.	A large distance.
C	High speed electron ejected from the nucleus as a neutron turns into a proton.	A few metres.

(a) (i) Identify the types of radiation **A**–**C**. (3)

A: ....

B: ....

C: ....

(ii) Rank the radiations in **(i)** from most ionising to least ionising. (2)

....

Most ionising   **Least ionising**

2 The equation below shows the decay of a radioactive element.

$$\underset{\textbf{A}}{\overset{238}{}}U \longrightarrow \underset{90}{\overset{\textbf{B}}{}}Th + \underset{\textbf{D}}{\overset{\textbf{C}}{}}\alpha$$

(a) What type of decay is shown by the equation? (1)

....

(b) Complete the missing parts (**A**–**D**) of the equation. (3)

A: ....

B: ....

C: ....

D: ....

(c) What effect does gamma emission have on the nucleus of a radioactive element? (1)

....

....

For more help on this topic, see Letts GCSE Combined Science Higher Revision Guide pages 222–223

1 Complete the following paragraph using some of the words below. (3)

| decay | irradiation | Sievert | contamination |
| radioactive | half-life | emit | radiation |

Radioactive .... is the unwanted presence of materials containing

radioactive atoms on other materials. This is a hazard due to the .... of

the contaminating atoms. The atoms will .... radiation so may be a hazard.

.... is the process of exposing an object to nuclear ....

without the object itself becoming ....

2 The graph below shows the decay of a radioactive element.

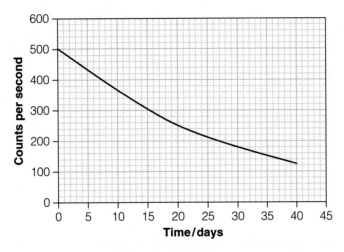

(a) (i) Estimate the half-life of this element. (1)

....

(ii) Explain how you arrived at your answer. (1)

....

(iii) Estimate the counts per second of the sample after 60 days. (2)

....

....

For more help on this topic, see Letts GCSE Combined Science Higher Revision Guide pages 224–225

GCSE
Combined Science
Paper 1: Biology 1

Higher
Time: 1 hour 15 minutes

You may use:
- a calculator
- a ruler.

Instructions

- Use black ink or black ball-point pen. Draw diagrams in pencil.
- Read each question carefully before you start to write your answer.
- Answer **all** questions in the spaces provided.
- Show your working in any calculator question and include units in your answer where appropriate.
- In questions marked with an asterisk (*****), marks will be awarded for your ability to structure your answer logically, showing how the points that you make are related or follow on from each other where appropriate.

Information

- The marks for each question are shown in brackets.
  Use this as a guide to how much time to spend on each question.
- The maximum mark for this paper is 70.
- Diagrams are not accurately drawn unless otherwise stated.

Name: ....

1 **(a)** Which substance is a product of anaerobic respiration in humans? Tick (✓) **one** box. [1]

Carbon dioxide ☐

Ethanol ☐

Glucose ☐

Lactic acid ☐

(b) Which substance is a product of aerobic respiration in plants? Tick (✓) **one** box. [1]

Carbon dioxide ☐

Ethanol ☐

Glucose ☐

Lactic acid ☐

2

Figure 1

(a) Niamh in **Figure 1** is training for a marathon. Every few days she runs a long distance. This builds up the number of mitochondria in her muscle cells.

What is the advantage for Niamh of having extra mitochondria in her muscle cells?
Tick (✓) **one** box. [1]

Her muscles become stronger. ☐

Her muscles can contract faster. ☐

Her muscles can release more energy. ☐

Her muscles can repair faster after injury. ☐

*(b) Bob in **Figure 2** has been running hard and has an oxygen debt. Describe what causes an oxygen debt after a session of vigorous exercise and how Bob can recover from its effects. **[3]**

Figure 2

....

....

....

....

*(c) Tariq is competing in a 10-mile running race. His heart rate and breathing rate increase. Describe how this helps his muscles during the race. **[3]**

....

....

....

3 Which of the following structures are in the correct size order, starting with the smallest? Tick (✓) **one** box.

[1]

cell, tissue, organ, system ☐

tissue, organ, system, cell ☐

cell, organ, system, tissue ☐

tissue, system, cell, organ ☐

4 Which part of the blood helps protect the body from pathogens? Tick (✓) **one** box.

[1]

Plasma ☐

Platelets ☐

Red blood cells ☐

White blood cells ☐

5 **(a)** From which section of the human heart is blood pumped to the lungs?
Tick (✓) **one** box. [1]

Left atrium ☐

Right atrium ☐

Left ventricle ☐

Right ventricle ☐

(b) Explain why the left side of the heart is more muscular than the right side. [1]

....

....

(c) The heart contains a number of valves.

Describe how the valves help the heart pump blood more effectively. [1]

....

....

(d) There are around 105 650 deaths in the UK from smoking-related causes each year. The number of deaths due to cardiovascular disease that is linked to smoking is estimated to be 22 100.

What percentage of deaths from smoking each year is due to cardiovascular disease? Show your working. Give your answer to two significant figures. [2]

....

(e) The chance of heart disease can be reduced by lowering alcohol intake and by not smoking. Write about **two** other choices that can be made to reduce the chance of heart disease. [2]

....

....

6 **(a)** Explain how skin defends the human body against disease. [2]

....

....

(b) Disease can be caused by bacteria. Bacteria multiply very quickly – the numbers of cholera bacteria can double every 20 minutes.

Ten cholera bacteria were kept in ideal conditions for growth. How many cholera bacteria were there after two hours? Show your working. **[2]**

....

....

(c) Explain why measles cannot be treated with antibiotics. **[2]**

....

....

(d) Figure 3 shows a photomicrograph of the bacterium *E. coli*, which is a common cause of stomach upsets. The magnification of the photo is × 5000.

Figure 3

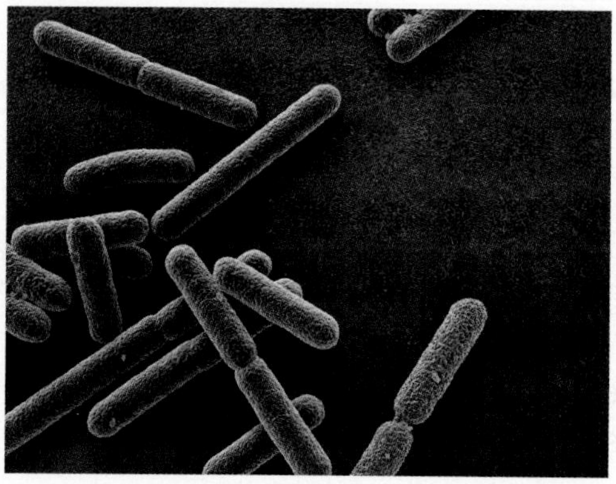

The average magnified length of the bacteria shown in **Figure 3** is 10 mm. Calculate the actual length of an *E. coli* bacterium in μm (1 μm = 1×10^{-3} mm). Show your working. **[2]**

....

....

(e) The spread of cholera is often a problem in temporary refugee camps. One way cholera is spread is through contaminated water.

What simple measures can be taken to reduce the spread of cholera in the camps? **[2]**

....

....

(f) If a disease is infectious, an epidemic can be prevented by vaccination.

Explain why a very high percentage of the population need to be vaccinated for the prevention to be successful. **[2]**

....

....

7 Which of the following is found in a plant cell but not in an animal cell?
Tick (✓) **one** box. **[1]**

Cell membrane ☐

Cytoplasm ☐

Nucleus ☐

Cell wall ☐

*8 Describe the aseptic method used to grow a colony of bacteria cells from pond water using a sterilised petri dish containing nutrient agar jelly. **[4]**

....

....

....

....

9 Human stem cell research could lead to new treatments for conditions such as diabetes.

(a) Outline **two** reasons why some people think that human stem cell research should not be allowed. **[2]**

....

....

(b) Look at **Figure 4**. What is the name of the process by which a stem cell becomes a new cell type? **[1]**

Figure 4

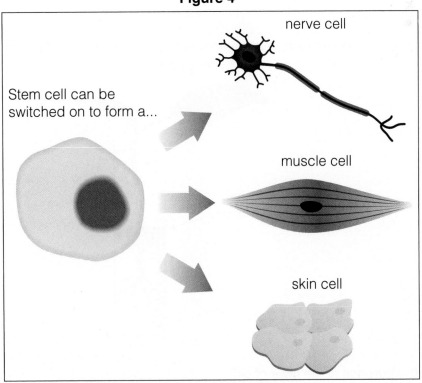

....

10 (a) Stomata are found on the underside of a plant leaf. During the day the stomata open to allow carbon dioxide into the leaf. Outline how the stomata open. **[2]**

....

....

(b) Carbon dioxide moves into a leaf by diffusion. Describe the process of diffusion. **[2]**

....

....

11 (a) Figure 5 shows a celery plant. The stalk of a celery plant contains xylem vessels.

Figure 5

Describe the structure of xylem vessels. **[2]**

....

....

(b) Describe how root cells are adapted to allow the efficient absorption of water. **[2]**

....

....

12 Figure 6 shows an experiment to investigate how water is taken up by a plant.

Figure 6

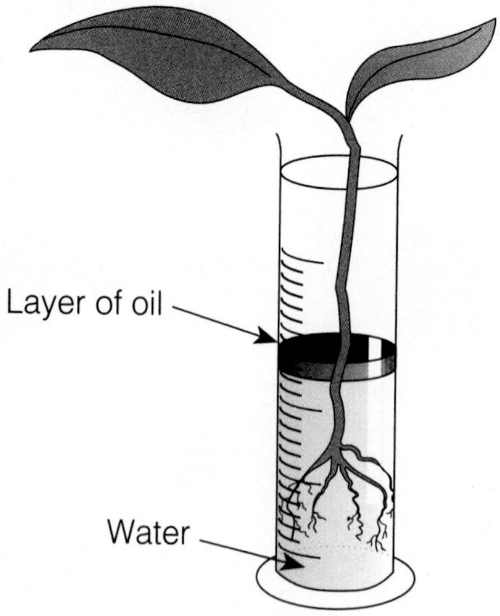

Layer of oil

Water

(a) What is the purpose of the layer of oil? **[1]**

....

(b) Table 1 shows the results of the experiment.

Table 1

Time (days)	0	1	2	3	4
Volume of water in cylinder (cm³)	50	47	43	42	40

Calculate the average water loss per day from the measuring cylinder in **Figure 6**.
Show your working. **[2]**

....

....

Average water loss: .... cm³ per day

13 Plants and animals use glucose for respiration. Plants also convert glucose into different substances.
Name **three** of these substances and give a reason why each one is important. **[3]**

....

....

....

14 **Figure 7** shows an experiment to demonstrate osmosis.

Figure 7

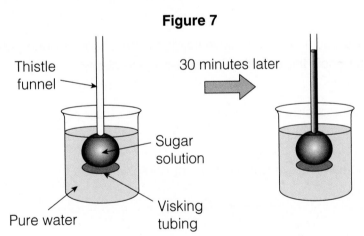

Explain why the volume of solution inside the thistle funnel has increased. **[3]**

....

....

....

*15 Black spot is a fungal disease that affects rose leaves (see **Figure 8**). Fatima has noticed that roses growing in areas with high air pollution are less affected by the disease. She thinks that regular exposure to acid rain is stopping the disease from developing.

Figure 8

Describe a simple experiment that Fatima could carry out to test her theory. [3]

....

....

....

....

16 **(a)** Thomas burns paraffin in a small stove inside his greenhouse during March and April.
How does this benefit the growth of his greenhouse plants? [2]

....

....

(b) Complete the balanced symbol equation for photosynthesis. [2]

$$6\,CO_2 + \text{.....................} \longrightarrow C_6H_{12}O_6 + \text{.....................}$$

*(c) Anoushka investigates the effect of light on photosynthesis at 25 °C.
She uses the apparatus shown in **Figure 9**.

Figure 9

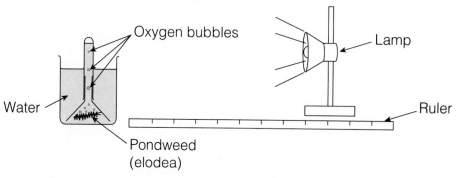

She moves the lamp and counts the bubbles at different distances from the pondweed. **Table 2** shows her results.

Table 2

Distance from lamp to pondweed (cm)	Number of bubbles counted in five minutes
10	241
20	124
30	60
40	36
50	24

Describe the pattern in the results in **Table 2** and predict how the pattern would change if the experiment was repeated at 35 °C and at 55 °C. Explain your answers. **[6]**

....

....

....

....

....

....

....

....

....

....

17 Outline why it is difficult to develop drugs that disable or destroy viruses. **[2]**

....

....

TOTAL FOR PAPER = 70 MARKS

GCSE

Combined Science

Paper 2: Biology 2

Higher

Time: 1 hour 15 minutes

You may use:

- a calculator
- a ruler.

Instructions

- Use black ink or black ball-point pen. Draw diagrams in pencil.
- Read each question carefully before you start to write your answer.
- Answer **all** questions in the spaces provided.
- Show your working in any calculator question and include units in your answer where appropriate.
- In questions marked with an asterisk (*), marks will be awarded for your ability to structure your answer logically, showing how the points that you make are related or follow on from each other where appropriate.

Information

- The marks for each question are shown in brackets.
  Use this as a guide to how much time to spend on each question.
- The maximum mark for this paper is 70.
- Diagrams are not accurately drawn unless otherwise stated.

Name: ....

1 **Figure 1** shows a simple food chain. The arrows represent the transfer of energy in the food chain.

Figure 1

Sun   **A**   Grass   **B**   Rabbit   **C**   Stoat   **D**   Fox

Which arrow shows the greatest transfer of energy?
Tick (✓) **one** box.   **[1]**

A   ☐

B   ☐

C   ☐

D   ☐

2 **Figure 2** shows a flock of seagulls. The number of seagulls in a flock is affected by both abiotic and biotic factors.

Figure 2

Which of the following is a biotic factor?
Tick (✓) **one** box.   **[1]**

Temperature   ☐

Rainfall   ☐

Disease   ☐

Ocean tides   ☐

3 Polar bears live in the Arctic. As **Figure 3** shows, they are often found on ice floes where they hunt seals.

Figure 3

Describe **two** adaptations that help polar bears survive in the arctic wilderness. **[2]**

....

....

4 An ecologist surveys a meadow to determine the number of field voles living there.

Humane traps are set and after 8 hours, 15 voles have been captured.

Each vole is marked by shaving a small section of fur from its back and is then released.

The traps are set again.

After a further 8 hours, 21 voles are captured. Four of these have shaved backs.

Use this formula to calculate the number of field voles in the meadow:

$$\frac{\text{number in first sample} \times \text{number in second sample}}{\text{number in second sample previously marked}}$$

Show your working. **[2]**

....

....

5 **Figure 4** shows a kakapo (*Strigops habroptilus*), which is a flightless parrot found only on a few small islands in New Zealand.

Figure 4

(a) Using the Linnaean system, what is the genus of the Kakapo? Tick (✓) **one** box. **[1]**

Animalia ☐

Aves ☐

Strigops ☐

habroptilus ☐

(b) The kakapo is a herbivore. It lives in woodland and is an excellent tree climber.

What part does the kakapo play in the woodland ecosystem? Tick (✓) **one** box. **[1]**

Decomposer ☐

Primary consumer ☐

Producer ☐

Secondary consumer ☐

6 The following text is one of the instructions on a packet of antibiotic tablets (see **Figure 5**):

Complete the prescribed course of treatment as directed by your doctor.

Figure 5

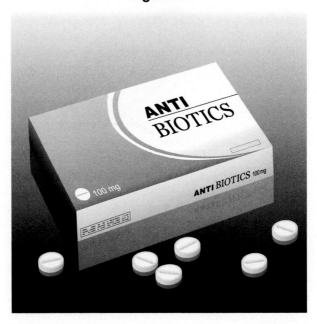

Explain why this instruction is important to stop bacteria becoming resistant to this antibiotic. **[3]**

....

....

....

7 Genetically engineered bacteria can be used to make human insulin to treat diabetes. The human genes to make insulin are inserted into the bacteria DNA.

Outline **three** reasons why bacteria are chosen for this process. **[3]**

....

....

....

8 In a recent newspaper article, the following statement appeared:

You can only get cancer if you have the wrong genes.

This is not true.

Write down **two** other factors that could be involved in the formation of cancer cells. **[2]**

....

....

....

9 Outline **two** reasons why Darwin's theory of evolution took many years to gain acceptance by the scientific community.

[2]

....

....

10 **Figure 6** shows some of the endocrine system.

Figure 6

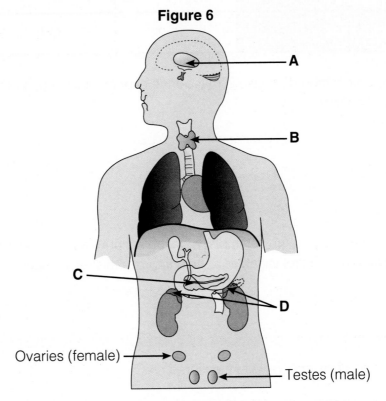

Which label shows the adrenal gland?
Tick (✓) **one** box.

[1]

A ☐

B ☐

C ☐

D ☐

11 Two students carried out an experiment to investigate the decay of grass cuttings.

- They put the cuttings in three different-coloured plastic bags – each bag had a few very small holes for ventilation.

- The bags were placed together at the edge of the school field.

- The bags were weighed at the start of the experiment and again three months later.

Table 1 shows the results.

Table 1

	Black bag (A)	Yellow bag (B)	White bag (C)
Mass of bag and grass cuttings at the start (g)	320.4	350.6	325.7
Mass of bag and grass cuttings after three months (g)	271.7	305.7	290.9
% decrease in mass	15.2	12.8	10.7

(a) Predict the percentage decrease in mass if the experiment is repeated with a dark brown bag. Explain your answer. **[3]**

....

....

....

(b) Putting the three bags in the same place on the school field is controlling a variable. Suggest **two** other control variables to help improve this experiment. **[2]**

1. ....

2. ....

12 The reaction times of six people were measured.

They put on headphones and were asked to push a button when they heard a sound. The button was connected to a timer.

Table 2 shows the results.

Table 2

Person	Gender	Reaction time in seconds				
		1	2	3	4	5
A	Male	0.26	0.25	0.27	0.25	0.27
B	Female	0.25	0.25	0.26	0.22	0.24
C	Male	0.31	1.43	0.32	0.29	0.32
D	Female	0.22	0.23	0.25	0.22	0.23
E	Male	0.27	0.31	0.30	0.28	0.26
F	Female	0.23	0.19	0.21	0.21	0.22

(a) Describe a pattern in these results. **[1]**

....

(b) Calculate the mean reaction time for person C, ignoring any outliers. Show your working. **[2]**

....

....

(c) When person D was concentrating on the test, someone touched her arm and she jumped. Her response was a reflex action. What are the **two** main features of a reflex action? **[2]**

....

....

*(d) A reflex action involves a 'message' travelling along a sensory neurone as an electrical impulse. The electrical impulse reaches a junction called a synapse.

Describe what happens at the synapse for the 'message' to continue its journey. **[4]**

....

....

....

....

13 **Figure 7** shows hormone levels in a woman's bloodstream during her menstrual cycle.

Figure 7

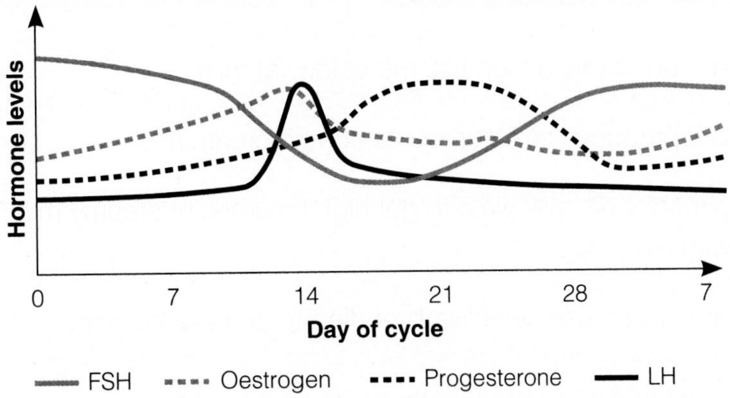

— FSH ···· Oestrogen ▪▪▪▪ Progesterone — LH

(a) The woman believes that her best chance of becoming pregnant is after day 21 of the cycle. Is she correct? Use the levels of hormones in **Figure 7** to explain your answer. **[2]**

....

....

....

(b) What is the role of progesterone in the menstrual cycle? **[2]**

....

....

(c) Explain how the hormones used in oral contraceptives prevent conception. **[2]**

....

....

(d) Fertility drugs have helped many couples have children, where previously they could not. However, the process is not without its problems.

Outline **two** problems with this type of treatment. **[2]**

....

....

*14 A farmer keeps two different herds of cows. The two types of cow are shown in **Figure 8**.

Figure 8

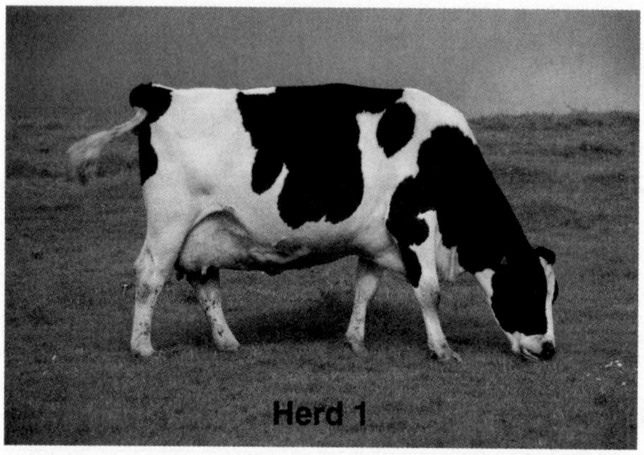

Herd 1

Herd 2

- Each cow in herd 1 produces a high volume of low fat milk.

- Each cow in herd 2 produces small amounts of high cream milk.

The farmer wants to modify his milk yield to get high volumes of creamy milk. He can sell the creamy milk to local ice-cream makers.

Describe how the farmer can use selective breeding to achieve his aim. **[4]**

....

....

....

....

15 **Figure 9** shows a sheepdog puppy. There are 39 pairs of chromosomes in the skin cell of a sheepdog puppy. Two of the chromosomes determine the sex of the puppy.

Figure 9

(a) How many chromosomes are there in the sperm cell of a sheepdog? Tick (✓) **one** box. **[1]**

39 ☐

76 ☐

78 ☐

80 ☐

(b) The number of skin cells increases as the puppy grows.

A skin cell grows and then divides into two skin cells. What is the name of this process?
Explain why each skin cell has the same number of chromosomes. **[4]**

....

....

....

....

16 Cystic fibrosis is a genetic disorder. Adam has the disorder but his mother and father did not show signs of the condition.

(a) Complete the Punnett diagram to show how this could occur. Use f as the allele for cystic fibrosis. **[2]**

(b) Before Adam was conceived, what was the probability that his parents would have a child with cystic fibrosis? **[1]**

....

(c) Which phrase describes the genotype of Adam's parents? **[1]**

....

*17 (a) Diabetes is a condition where the body cannot control the level of glucose in the blood.

Describe how the human body normally controls the level of glucose in the blood and how people with type 1 and type 2 diabetes maintain the correct levels of glucose in the blood. **[6]**

....

....

....

....

....

....

....

....

....

(b) Diabetes can adversely affect the kidneys.

Kidneys remove urea from the bloodstream. Outline what other jobs the kidneys do. **[2]**

....

....

(c) A person with kidney disease must control their diet to reduce the kidneys' workload.

Explain in detail why the amount of protein in the diet must be limited. **[4]**

....

....

....

....

....

18 Large areas of Amazon rainforest are cleared to provide land for farming every year.
Explain how deforestation is linked to global warming. **[2]**

....

....

....

19 Write down **two** reasons why biodiversity is important. **[2]**

....

....

TOTAL FOR PAPER = 70 MARKS

GCSE
Combined Science
Paper 3: Chemistry 1

Higher
Time: 1 hour 15 minutes

You may use:

- a calculator
- a ruler.

Instructions

- Use black ink or black ball-point pen. Draw diagrams in pencil.
- Read each question carefully before you start to write your answer.
- Answer **all** questions in the spaces provided.
- Show your working in any calculator question and include units in your answer where appropriate.
- In questions marked with an asterisk (*****), marks will be awarded for your ability to structure your answer logically, showing how the points that you make are related or follow on from each other where appropriate.

Information

- The marks for each question are shown in brackets.
  Use this as a guide to how much time to spend on each question.
- The maximum mark for this paper is 70.
- Diagrams are not accurately drawn unless otherwise stated.

Name: _____

1 The periodic table lists all known elements.

(a) Which statement about the periodic table is correct?
Tick (✓) **one** box. **[1]**

Each row begins with elements with one outer electron. ☐

The columns are called periods. ☐

The elements are arranged in mass number order. ☐

The metallic elements are on the right. ☐

(b) Which of these statements about the elements in Group 0 is correct?
Tick (✓) **one** box. **[1]**

They are all liquids at room temperature. ☐

Their boiling points increase as you go down the group. ☐

They have very high melting points. ☐

Their molecules are made from pairs of atoms. ☐

(c) Element X is a solid with a low melting point. When it reacts it forms covalent bonds with other elements or it forms negative ions.

Put an X where you would expect to find element X on the periodic table. **[1]**

2 **(a)** Which statement explains why group 1 elements are known as the alkali metals?
Tick (✓) **one** box. **[1]**

They are tested with an alkali to show they are reactive. ☐

They are in the first column in the periodic table. ☐

They all react strongly with alkalis. ☐

They make an alkali when reacted with water. ☐

(b) Sodium is below lithium in group 1 of the periodic table.

Explain why sodium reacts more vigorously with water than lithium. **[2]**

....

....

(c) Figure 1 shows the arrangement of particles in a metal.

Figure 1

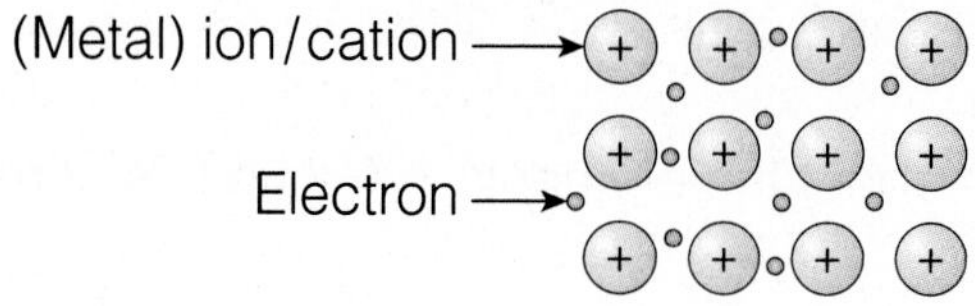

(Metal) ion / cation ⟶

Electron ⟶

Explain how metals conduct electricity. **[2]**

....

....

3 **(a)** Graphite is commonly used as a lubricant in machines that operate at high temperatures.

Which properties of graphite explain why it is suitable for this use?
Tick (✓) **one** box. **[1]**

Electrical conductor and high melting point ☐

Good heat and electrical conductor ☐

Good heat conductor and slippery ☐

High melting point and slippery ☐

(b) Graphene is a form of carbon. It is formed of a sheet of carbon atoms, one atom thick.

A graphene sheet has a thickness of 3.4×10^{-8} cm. Calculate the area covered by 1 cm³ of graphene.
Tick (✓) **one** box. **[1]**

3.4×10^{8} cm² ☐

2.9×10^{7} cm² ☐

2.9×10^{-7} cm² ☐

3.4×10^{-8} cm² ☐

(c) Carbon nanotubes are cylindrical fullerenes.

Outline two important physical properties of nanotubes. **[2]**

....

....

(d) Diamond has a tetrahedral structure, as shown in **Figure 2**.

Figure 2

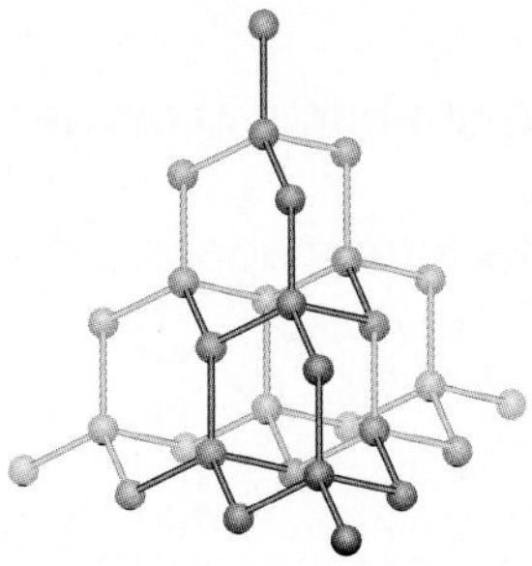

Explain why diamond has a very high melting point and why, unlike graphite, it does not conduct electricity. **[3]**

....

....

....

4 **(a)** Carbon dioxide is made by the thermal decomposition of copper(II) carbonate (see **Figure 3**). Copper(II) oxide is also made.

Figure 3

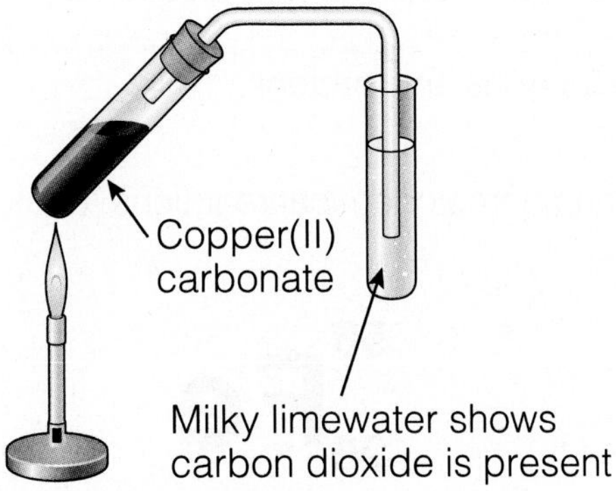

Copper(II) carbonate

Milky limewater shows carbon dioxide is present

Write the word equation for the decomposition reaction. **[1]**

....

(b) Calculate the mass of carbon dioxide made when 12.35 g of copper(II) carbonate is heated to make 7.95 g of copper(II) oxide. Show your working. **[2]**

....

....

5 **Figure 4** shows the electronic structure of an oxygen atom and a magnesium atom.

Figure 4

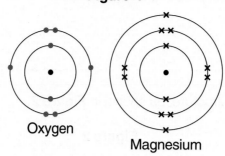

Oxygen

Magnesium

Electrons are transferred when magnesium burns in oxygen to produce magnesium oxide.

(a) Describe how the magnesium atoms form magnesium ions and the oxygen atoms form
oxide ions. **[2]**

....

....

(b) Give the charge of each ion. **[2]**

....

6 **(a)** Which of these statements about a neutral atom is always correct?
Tick (✓) **one** box. **[1]**

It has the same number of electrons and neutrons. ☐

It has the same number of protons and neutrons. ☐

It has the same number of protons, neutrons and electrons. ☐

It has the same number of electrons and protons. ☐

(b) (i) Fe^{2+} ions are formed during some chemical reactions. Look at the information given below and
then complete **Table 1**. **[1]**

$$^{56}_{26}Fe$$

Table 1

Number of protons in the ion	
Number of neutrons in the ion	
Number of electrons in the ion	

*(ii) Explain how you worked out each of the three numbers. **[3]**

Number of protons:

....

....

Number of neutrons:

....

....

Number of electrons:

....

....

7 Gold metal can be rolled into very thin sheets called gold leaf.

The radius of a gold atom is 1.5×10^{-10} m.

Gold leaf has a typical thickness of 1.2×10^{-6} m.

Calculate how many gold atoms are packed on top of each other to achieve this thickness. **[2]**

....

....

....

8 **(a)** Acids react with bases to form salts and water. Which pair of reactants can be used to prepare copper sulfate?
Tick (✓) **one** box. **[1]**

Copper and sulfuric acid ☐

Copper hydroxide and nitric acid ☐

Copper oxide and sulfuric acid ☐

Copper oxide and hydrochloric acid ☐

(b) Josh put a sample of potassium hydroxide solution into a beaker. He measured the pH. Then he slowly added dilute nitric acid until no further reaction took place.

How would the pH of the solution in the beaker change?
Tick (✓) **one** box. [1]

The pH would start high and decrease to below 7. ☐

The solution would change to a pH of 7. ☐

The pH would stay the same. ☐

The pH would start low and increase to above 7. ☐

(c) Which of the following 0.1 mol/dm³ acid solutions has the lowest pH?
Tick (✓) **one** box. [1]

Carbonic acid ☐

Citric acid ☐

Ethanoic acid ☐

Nitric acid ☐

(d) An acid–base reaction was completed between hydrochloric acid (HCl) and calcium oxide (CaO) to make calcium chloride ($CaCl_2$).

This is the equation for the reaction: $2HCl + CaO \rightarrow CaCl_2 + H_2O$

An excess of solid calcium oxide was added to the acid.

Calculate the minimum mass of calcium oxide needed to make 5.55 g of calcium chloride.
Show your working. [4]

....

....

....

....

9 Iron(III) oxide is roasted with carbon (coke) in a blast furnace to produce iron. In one of the reactions in the furnace, carbon reacts with oxygen in the air to make carbon monoxide.

Carbon monoxide (CO) then reacts with the iron(III) oxide (Fe_2O_3) to make iron (Fe). The other product is carbon dioxide.

(a) Write a balanced symbol equation for the reaction of carbon monoxide with iron(III) oxide. [2]

....

(b) Heating a metal oxide with carbon is a common method used to extract the metal.

Explain why copper can be extracted from copper oxide but aluminium cannot be extracted from its oxide by this method. **[2]**

....

....

(c) Aluminium is extracted by the electrolysis of molten aluminium oxide (Al_2O_3).

Write the ionic half equation for the reaction at each electrode. **[2]**

Cathode: ....

Anode: ....

10 Iris measured 15 cm³ of water into a test tube, as shown in **Figure 5**.

Figure 5

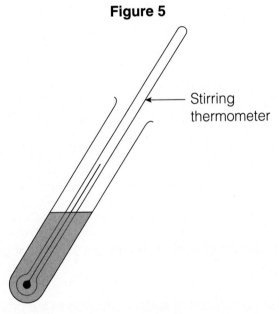

She measured the temperature of the water and added 2 g of a solid. She stirred until there was no further temperature change. She repeated the experiment with other solids.

(a) Complete the results table (**Table 2**). **[2]**

Table 2

Solid	Start temperature (°C)	End temperature (°C)	Temperature change (°C)
Ammonium chloride	15	9	−6
Potassium hydroxide	16	29	+13
Ammonium nitrate	18	4	....
Sodium hydroxide	17	35	....

*(b) Which of the solids had the largest endothermic energy change? Explain your answer.   [3]

....

....

....

*11 Our understanding of the model of the atom has developed from the work of a number of scientists, starting from Dalton's theory that an atom was a solid sphere.

Outline how our understanding of the atom has changed. Link the key scientists with the improvements they made to our understanding.   [6]

....

....

....

....

....

....

....

....

....

....

*12 (a) Finlay added an aqueous solution of sodium iodide to a solution of bromine. The colour changed from orange to deep brown.

Finlay then added an aqueous solution of sodium chloride to the bromine solution. The orange colour did not change.

Explain these observations.   [4]

....

....

....

....

(b) Chlorine is composed of diatomic molecules, Cl_2.

Draw a dot-and-cross diagram to show the bonding in a chlorine molecule. You should only show the outer shell electrons in your diagram. **[2]**

(c) Chlorine and iodine are both in group 7 of the periodic table.

Explain why chlorine is a gas and iodine is a solid at room temperature. **[2]**

....

....

(d) Explain why solid iodine does not conduct electricity. **[1]**

....

13 **(a)** When 1 mole of carbon burns completely, 393 kJ of energy is released.

$$C_{(s)} + O_{2(g)} \rightarrow CO_{2(g)}$$

The relative atomic mass (A_r) of carbon = 12.

Calculate the energy released when 14.4 g of carbon is burned. Show your working. **[2]**

....

....

....

*(b) Energy is released when carbon burns. Use ideas about bond making and bond breaking to explain why. **[3]**

....

....

....

....

(c) Amy measured the energy released by reacting hydrochloric acid with sodium hydroxide solution. Both solutions had the same concentration.

This was the method used.

1 Measure 25 cm³ sodium hydroxide solution using a 100 cm³ measuring cylinder.

2 Pour the sodium hydroxide solution into a 250 cm³ beaker.

3 Use the 100 cm³ measuring cylinder to measure 25 cm³ hydrochloric acid.

4 Pour the acid into the sodium hydroxide in the beaker.

5 Measure the start temperature with a thermometer.

6 After one minute, measure the final temperature.

This method gave a poor result. Suggest three improvements to the method. **[3]**

1: ....

....

2: ....

....

3: ....

....

TOTAL FOR PAPER = 70 MARKS

GCSE
Combined Science

Paper 4: Chemistry 2

Higher

Time: 1 hour 15 minutes

You may use:

- a calculator
- a ruler.

Instructions

- Use black ink or black ball-point pen. Draw diagrams in pencil.
- Read each question carefully before you start to write your answer.
- Answer **all** questions in the spaces provided.
- Show your working in any calculator question and include units in your answer where appropriate.
- In questions marked with an asterisk (*), marks will be awarded for your ability to structure your answer logically, showing how the points that you make are related or follow on from each other where appropriate.

Information

- The marks for each question are shown in brackets.
  Use this as a guide to how much time to spend on each question.
- The maximum mark for this paper is 70.
- Diagrams are not accurately drawn unless otherwise stated.

Name: ....

1 **(a)** Which one of the following gases will bleach damp litmus paper?
Tick (✓) **one** box. **[1]**

Carbon dioxide ☐

Chlorine ☐

Hydrogen ☐

Methane ☐

(b) Which one of the following gases is made when ethanoic acid reacts with calcium carbonate?
Tick (✓) **one** box. **[1]**

Carbon dioxide ☐

Chlorine ☐

Hydrogen ☐

Methane ☐

2 The apparatus used in the laboratory for cracking long-chain hydrocarbons is shown in **Figure 1**.

Figure 1

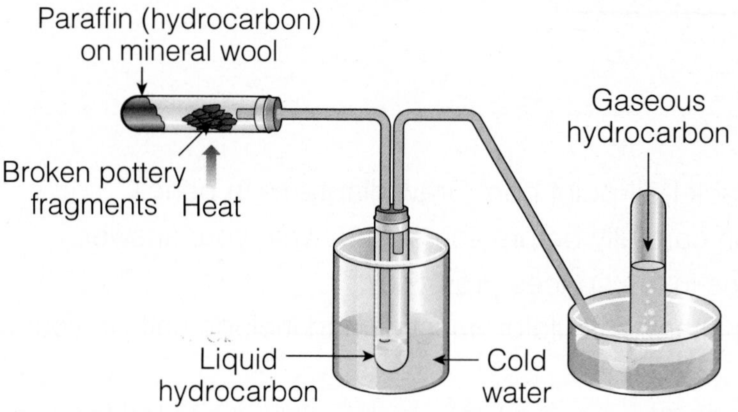

Paraffin (hydrocarbon) on mineral wool

Gaseous hydrocarbon

Broken pottery fragments   Heat

Liquid hydrocarbon — Cold water

(a) Explain what is meant by **cracking** long-chain hydrocarbons. **[2]**

....

....

(b) What is the purpose of the broken pottery fragments? **[1]**

....

(c) The paraffin on the mineral wool has the formula $C_{16}H_{34}$. The gaseous product is ethene (C_2H_4) and the liquid hydrocarbon is decane ($C_{10}H_{22}$).

Construct a balanced symbol equation for the reaction. **[2]**

....

*3 Explain how the difference in strength of intermolecular forces between hydrocarbons allows them to be separated by fractional distillation. [3]

....

....

....

4 Which of the following hydrocarbons is the most flammable?
Tick (✓) **one** box. [1]

C_8H_{18} ☐

$C_{11}H_{24}$ ☐

C_5H_{12} ☐

$C_{14}H_{30}$ ☐

5 Five students each have a test tube containing $10\,cm^3$ of hydrochloric acid of the same concentration. They each have a different-sized strip of magnesium ribbon.

They drop the magnesium into the acid and time how long it takes for the fizzing to stop.

Table 1 shows the results of the experiment.

Table 1

Student	Iram	Alex	Dylan	Georgie	Noah
Time (s)	246	258	204	300	272

(a) Which student had the fastest reaction? [1]

....

(b) Georgie noticed that there was some magnesium left in the test tube when the fizzing stopped. In all the other test tubes, there was no magnesium left. Explain these two observations. [2]

....

....

....

(c) (i) Georgie repeated her experiment.

This time she measured the volume of gas made with a gas syringe. She measured 94.5 cm³ of gas made in 225 seconds.

Calculate the mean rate of reaction. Show your working and include the unit in your answer. **[3]**

....

....

....

(ii) Georgie repeated her experiment again. This time she used double the volume of acid and double the amount of magnesium.

Predict what happened to the amount of gas made in the reaction. Explain your answer. **[2]**

....

....

....

6 **Figure 2** shows the result of a chromatography experiment on an unknown black ink.

Figure 2

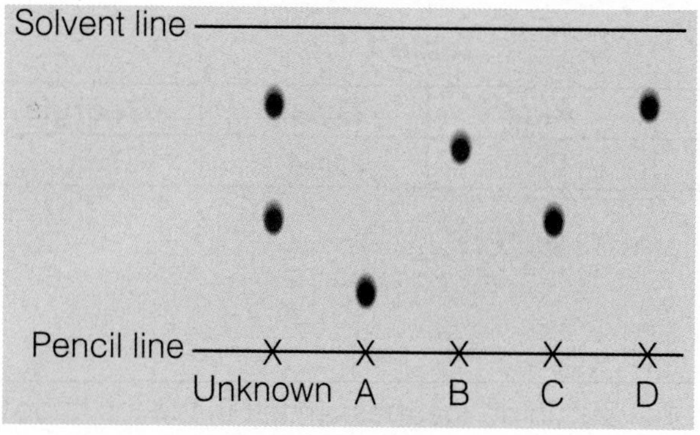

(a) Which inks does the unknown ink in **Figure 2** contain?
Tick (✓) **one** box. **[1]**

A and B ☐

A and C ☐

B and D ☐

C and D ☐

(b) The R$_f$ value of ink B is 0.86. The solvent line moved 7.91 cm from the pencil line.

Calculate how far ink B moved up the paper. Show your working. Give your answer to an appropriate number of significant figures. **[2]**

....

....

7 **(a)** Which of the following molecules has the formula C_4H_{10}? Tick (✓) **one** box. [1]

Structure 1:
$$H-\overset{\overset{\displaystyle H}{|}}{\underset{\underset{\displaystyle H}{|}}{C}}-\overset{\overset{\displaystyle H}{|}}{\underset{\underset{\displaystyle H}{|}}{C}}-\overset{\overset{\displaystyle O}{|}}{\underset{\underset{\displaystyle H}{|}}{C}}-\overset{\overset{\displaystyle H}{|}}{\underset{\underset{\displaystyle H}{|}}{C}}-H$$

☐

Structure 2:
$$\begin{array}{c}H\\ \diagdown \\ H \diagup\end{array} C=\overset{\overset{\displaystyle H}{|}}{C}-\overset{\overset{\displaystyle H}{|}}{\underset{\underset{\displaystyle H}{|}}{C}}-\overset{\overset{\displaystyle H}{|}}{\underset{\underset{\displaystyle H}{|}}{C}}-H$$

☐

Structure 3:
$$H-\overset{\overset{\displaystyle H}{|}}{\underset{\underset{\displaystyle H}{|}}{C}}-\overset{\overset{\displaystyle H-\overset{\overset{\displaystyle H}{|}}{\underset{\underset{\displaystyle H}{}}{C}}-H}{|}}{\underset{\underset{\displaystyle H}{|}}{C}}-\overset{\overset{\displaystyle H}{|}}{\underset{\underset{\displaystyle H}{|}}{C}}-H$$

☐

Structure 4:
$$\begin{array}{c}H\\ \diagdown \\ H \diagup\end{array} C=\overset{\overset{\displaystyle H-\overset{\overset{\displaystyle H}{|}}{\underset{\underset{\displaystyle H}{}}{C}}-H}{|}}{C}-\overset{\overset{\displaystyle H}{|}}{\underset{\underset{\displaystyle H}{|}}{C}}-H$$

☐

(b) Write the balanced symbol equation for the complete combustion of C_4H_{10}. [2]

....

8 A nine carat wedding ring weighs 4.5 g. What is the weight of pure gold in the ring?
Pure gold is 24 carats.
Tick (✓) **one** box. [1]

0.50 g ☐

0.90 g ☐

1.69 g ☐

4.05 g ☐

9 **(a)** These statements describe the process by which the Earth's atmosphere has changed.

    A Oceans formed as the temperature at the surface fell below 100°C.

    B Photosynthesis released oxygen into the atmosphere and used up carbon dioxide.

    C Nitrifying bacteria used up ammonia and released nitrogen.

    D Hot volcanic earth released carbon dioxide and ammonia into the atmosphere.

Put each letter in the correct box to show the order that scientists now believe the atmosphere developed. [2]

☐ → ☐ → ☐ → ☐

(b) How the Earth's atmosphere evolved is a theory. What is a theory? [2]

....

....

(c) Explain why the way that the Earth's atmosphere evolved can only be a theory. [1]

....

....

(d) Which pie chart shows the composition of the Earth's atmosphere today? Tick (✓) **one** box.   **[1]**

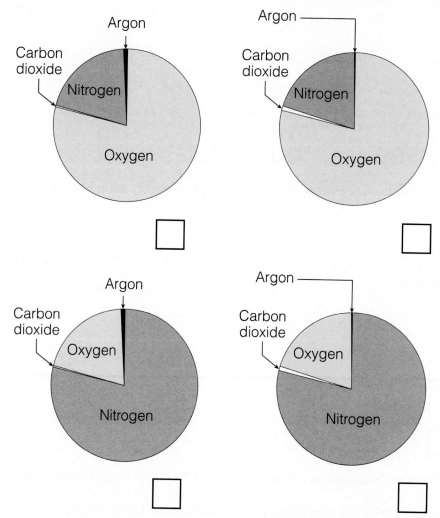

10 Meg carried out a rate of reaction experiment by reacting hydrochloric acid with sodium thiosulfate solution. A yellow precipitate of sulfur formed.

As shown in **Figure 3**, the reaction was followed by timing how long it took a cross drawn under a flask to disappear.

Figure 3

All reactions were carried out at 25°C. **Table 2** shows the results.

Table 2

Concentration of acid (mol/dm³)	Time taken for cross to disappear (s)
0.1	60
0.2	40
0.4	24
0.6	13
0.8	8
1.0	4

(a) Plot the results on the graph paper in **Figure 4**.   [3]

Figure 4

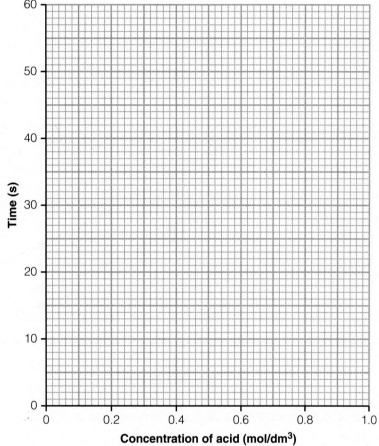

*(b) Describe and explain how the rate of reaction changes as the concentration of acid changes.   [3]

....

....

....

(c) The experiment was repeated at 35°C. Predict how the reaction times would change.   [1]

....

11 Explain the purpose of desalination and chlorination in making safe drinking water. **[2]**

Desalination:

....

....

Chlorination:

....

....

12 **Figure 5** shows how the yield of the Haber process changes with different conditions.

Figure 5

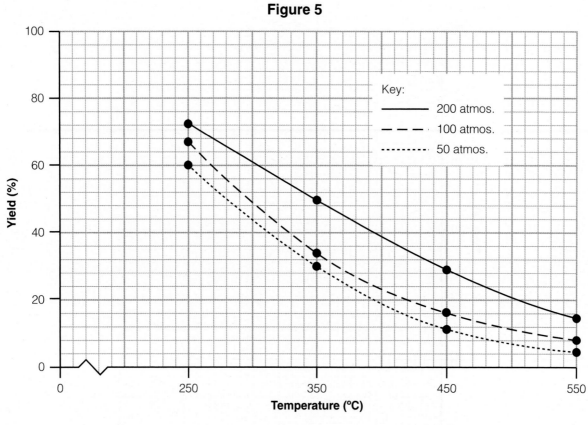

(a) Write down the yield at 200 atmospheres and at a temperature of 350°C. **[1]**

....

(b) Describe what happens to the yield as the temperature is increased. **[1]**

....

(c) This is the equation for the reaction: $N_{2(g)} + 3H_{2(g)} \rightleftharpoons 2NH_{3(g)}$

Use the equation to explain why the yield increases with increased pressure. **[2]**

....

....

....

(d) Very high yields of ammonia can be achieved at a pressure of 500 atmospheres.

Explain why the normal operating pressure for the Haber process is much lower.   **[2]**

....

....

....

13 **Figure 6** shows the results of an investigation into the reaction of zinc metal with hydrochloric acid.

Experiments A and B used 2 g of zinc: one experiment used zinc powder and the other used zinc granules.

Figure 6

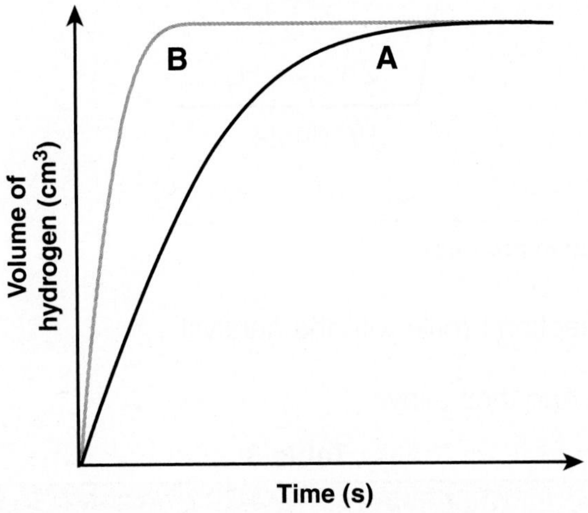

*(a) Which line on the graph represents the reaction with powdered zinc?
Explain your answer using the idea of reacting particles.   **[4]**

....

....

....

....

(b) Copper ions (Cu^{2+}) act as a catalyst for the reaction.

Figure 7 shows the reaction profile without a catalyst.

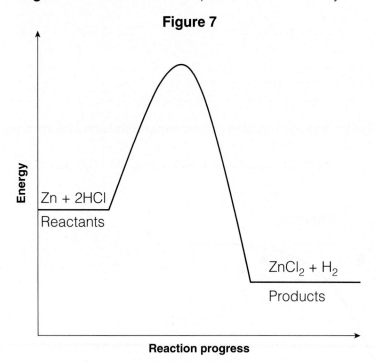

Figure 7

Draw on the graph the reaction profile with the catalyst. **[1]**

* 14 **Table 3** shows some metals and their alloys.

Table 3

	Order of hardness	Density (g/cm³)	Melting point (°C)	Order of strength
Copper	5	8.9	1083	5
Brass (alloy of copper)	3	8.6	920	3
Iron	2	7.9	1538	2
Steel (alloy of iron)	1	7.8	1420	1
Lead	6	11.3	327	6
Solder (alloy of lead)	5	9.6	170	4

Use the data in the table to outline what alloying does to the properties of pure metals. **[4]**

....

....

....

....

15 Look at the displayed formula of propene in **Figure 8**.

Figure 8

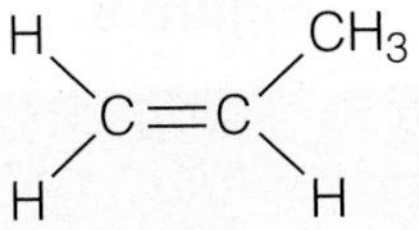

$$H_2C=CH-CH_3$$

Propene molecules can join together to form the addition polymer, poly(propene).

Draw a diagram to show the structure of the polymer.   **[3]**

16 Sulfuric acid is manufactured in the contact process.

In one of the reactions in the process, sulfur dioxide is converted to sulfur trioxide:

$$2SO_{2(g)} + O_{2(g)} \rightleftharpoons 2SO_{3(g)}$$

This is a reversible reaction and it will reach a position of equilibrium.

(a) Describe how the reaction reaches equilibrium from the start. Use ideas about rate of reaction.   **[2]**

....

....

(b) Predict and explain the effect of reducing the pressure on the position of equilibrium for this
reaction.   **[2]**

....

....

(c) What is the effect of using a catalyst on the position of equilibrium in this reaction?   **[1]**

....

(d) Outline the extra information you need to determine the effect on the equilibrium position of
increasing the temperature of the reaction mixture.   **[1]**

....

17 Catalytic converters are fitted to modern cars to reduce carbon monoxide and nitrogen dioxide emissions formed during the combustion of fuel (see **Figure 9**).

Figure 9

*(a) Describe how nitrogen dioxide gas is formed by the car. [3]

....

....

....

(b) Sulfur dioxide can also be formed from burning fuels such as petrol or diesel. Catalytic converters cannot reduce sulfur dioxide emissions.

(i) Give one reason why sulfur dioxide is an atmospheric pollutant. [1]

....

(ii) Suggest how emissions of sulfur dioxide can be reduced from cars that use diesel and petrol. [1]

....

TOTAL FOR PAPER = 70 MARKS

GCSE
Combined Science
Paper 5: Physics 1

Higher
Time: 1 hour 15 minutes

You may use:
- a calculator
- a ruler.

Instructions
- Use black ink or black ball-point pen. Draw diagrams in pencil.
- Read each question carefully before you start to write your answer.
- Answer **all** questions in the spaces provided.
- Show your working in any calculator question and include units in your answer where appropriate.
- In questions marked with an asterisk (*), marks will be awarded for your ability to structure your answer logically, showing how the points that you make are related or follow on from each other where appropriate.

Information
- The marks for each question are shown in brackets.
  Use this as a guide to how much time to spend on each question.
- The maximum mark for this paper is 70.
- Diagrams are not accurately drawn unless otherwise stated.

Name: ....

1 **(a)** Which of the following circuit diagram symbols represents a thermistor?
Tick (✓) **one** box. [1]

(b) The diagrams below show four circuits. Each circuit has two **identical** bulbs connected in parallel.

In which circuit will the ammeters have the same reading?
Tick (✓) **one** box. [1]

(c) A torch bulb has a resistance of 1200 Ω. The bulb operates when the current through it is 0.005 A.
How many 1.5 V batteries will the torch need to operate? [3]

....

....

....

*(d) Alex and Louise want to find out the identity of the mystery electrical component contained in Box Z. They connect it in the circuit shown in **Figure 1**.

Figure 1

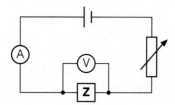

They change the potential difference, measure the current and plot the results (see **Figure 2**).

Figure 2

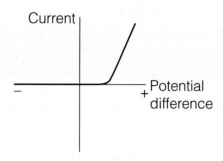

Write down the name of the mystery component in Box Z.
Explain your answer using ideas about resistance. [3]

....

....

....

2 **(a)** Complete **Table 1** to show the atomic structures of the three isotopes of carbon. [3]

Table 1

Isotope	Number of protons	Number of neutrons	Number of electrons
Carbon-12	6	6	6
Carbon-13			
Carbon-14			

(b) Carbon-14 is an unstable isotope. It undergoes beta decay to form nitrogen-14.

Describe what happens in the nucleus of an atom to form a beta particle. [2]

....

....

(c) The half-life for the decay of carbon-14 is 5730 years.

Explain what is meant by the term half-life. [1]

....

....

(d) The very small amount of carbon-14 in a plant remains constant until the plant dies. The amount of carbon-14 then falls steadily as the plant decays. The amount of carbon-12 stays the same.

The plant material can be dated by measuring the ratio of carbon-14 to carbon-12.

Complete **Table 2**. [1]

Table 2

Number of half-lives	Time after death of organism in years	$^{14}C : {}^{12}C$ ratio / 10^{-12}
0	0	1.000
1	5730	0.500
2	11 460	0.250
3	....	0.125

(e) Use the data from **Table 2** to plot a decay curve for carbon-14 (see **Figure 3**). [2]

Figure 3

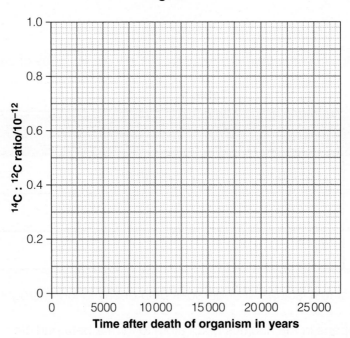

(f) A sample of wood taken from an old shipwreck had a carbon-14 : carbon-12 ratio / 10^{-12} of 0.3.

Use **Figure 3** to estimate the age of the wood. [1]

....

3 **(a)** Look at **Figure 4**. Which set of conditions would you find on the label of a microwave oven that operates with the mains electricity supply in the UK?

Tick (✓) **one** box. [1]

Figure 4

50 Hz, 230 V, ac ☐

50 Hz, 230 V, dc ☐

60 Hz, 230 V, dc ☐

60 Hz, 230 V, ac ☐

(b) **Figure 5** shows a symbol found on hairdryers.

Figure 5

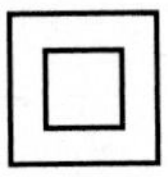

The symbol shows that the hairdryer is double insulated. What does this mean?

Tick (✓) **one** box. [1]

It has a moulded plastic plug. ☐

It does not need an earth wire. ☐

The case has metal parts on the outside. ☐

The dryer has double plastic coating on the wires. ☐

4 Look at the diagram of a three-pin plug in **Figure 6**.

Figure 6

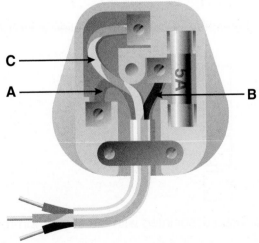

(a) Complete **Table 3**. [2]

Wire	Name	Colour
A	....	Blue
B	Live	....
C	....	Green and yellow stripes

(b) Give **two** reasons why wire C is connected to some domestic appliances. [2]

....

....

....

*(c) The plug is connected to an electric drill and contains a 5 A fuse.

The electric drill has a power rating of 960W.

Calculate the value of the current passing through the fuse when the drill is operating normally at 230 V. Show all your working. Give your answer to two significant figures. [3]

....

....

....

Answer: .... A

5 **(a)** Which of the following energy resources is non-renewable?
Tick (✓) **one** box. [1]

Bio-fuel ☐

Coal ☐

Hydro-electricity ☐

Wind ☐

(b) Which of the following energy resources can be most relied on to give a constant supply of energy?
Tick (✓) **one** box. [1]

Geothermal ☐

Tidal ☐

Wave ☐

Wind ☐

(c) (i) State **one** environmental issue associated with the use of non-renewable fuels. [1]

....

(ii) Give **two** reasons why we still use non-renewable fuels. [2]

....

....

6 Look at the picture of the roller coaster in **Figure 7**.

Figure 7

(a) At what position have the roller coaster cars got maximum kinetic energy?
Tick (✓) **one** box.   **[1]**

Position 1 ☐

Position 2 ☐

Position 3 ☐

Position 4 ☐

(b) Describe the energy changes that take place as the roller coaster cars travel from position 1 to
position 4.   **[2]**

....

....

....

*(c) Explain why the next 'hill' on the roller coaster after position 4 has to be lower than position 1.   **[3]**

....

....

....

7 **Figure 8** shows an electric screwdriver. This question is about motors used in electric screwdrivers.

Figure 8

Table 4

Electric motor	Input electrical power (W)	Useful output power (W)	Wasted power (W)
A	209	220	11
B	172	160	12
C	235	223	12
D	205	193	12
E	242	....	20

(a) (i) Complete **Table 4**. [1]

(ii) Name the motor with the largest useful power output. .... [1]

(b) Describe how the electrical power is used and suggest how some power is wasted. [2]

....

....

....

(c) Calculate the efficiency of electric motor B. [2]

....

....

....

8 **(a)** Which of the following vehicles has the greatest kinetic energy?
Tick (✓) **one** box. [1]

A van with a mass of 1500 kg travelling at 11 m/s ☐

A car with a mass of 1400 kg travelling at 12 m/s ☐

A van with a mass of 1200 kg travelling at 12 m/s ☐

A car with a mass of 1100 kg travelling at 13 m/s ☐

(b) Figure 9 shows an electric motor lifting a load.

Figure 9

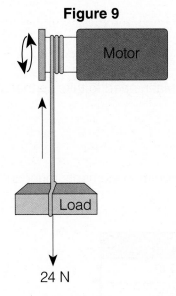

24 N

How much work (in joules) is done when the load is lifted through a height of 2.5 m?
Show all your working. **[2]**

....

....

(c) It takes four seconds to lift the load through 2.5 m. Calculate the power of the motor.
Show all your working. **[2]**

....

....

9 | **Figure 10** shows some meteorites that Mahri is investigating.

Figure 10

(a) Draw a labelled diagram to describe how she can measure the volume of each meteorite. **[2]**

(b) Mahri measures the volume of each meteorite three times. How does this improve her results? **[2]**

....

....

(c) What other measurement does Mahri need to take so that she can calculate the density of the meteorites? **[1]**

....

(d) Table 5 shows the densities of common elements found in meteorites.

Table 5

Element	Density kg/m³
Aluminium	2712
Iron	7850
Magnesium	1738
Nickel	8908
Silicon	2328

The mean density of the three meteorites was 7965 kg/m³. Analysis indicated that two elements were present. Iron formed 90% of the composition. What was the element making up the remaining 10%? Tick (✓) **one** box. **[1]**

Aluminium ☐

Magnesium ☐

Nickel ☐

Silicon ☐

10 Look at **Table 6**, which shows the specific heat capacity of four metals.

Table 6

Metal	Aluminium	Copper	Iron	Lead
Specific heat capacity J/kg °C	900	490	390	130

(a) If you were given a 2 kg block of each metal, which metal would take the least amount of energy to raise its temperature from 20°C to 25°C?

Tick (✓) **one** box.   [1]

Aluminium ☐

Copper ☐

Iron ☐

Lead ☐

(b) As shown in **Figure 11**, a 500 g block of copper was heated with a 100 W electric immersion heater for 85 seconds.

Figure 11

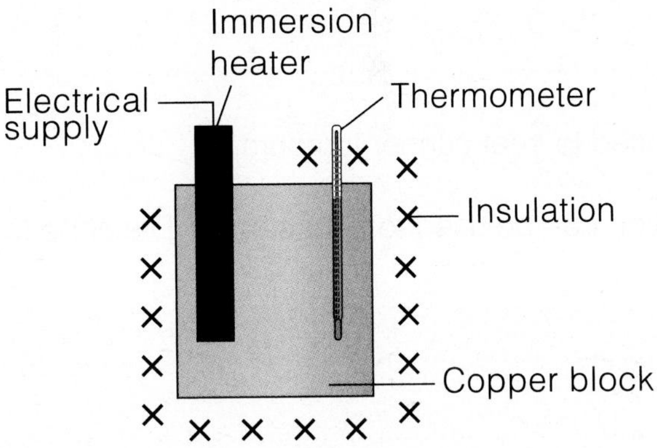

Calculate the amount of energy in joules supplied to the copper block. Show all your working.   [2]

....

....

....

(c) In a repeat experiment, the copper block started at a temperature of 22°C and was supplied with 9065 J of energy. What was its final temperature? Show all your working.   [3]

change in thermal energy = mass × specific heat capacity × temperature change

....

....

....

*11 A well-insulated beaker contains 200 g of water at 20°C. This will release 16 800 J of energy to cool down to 0°C.

60 g of ice at 0°C is added to the water.

If no heat energy is lost or gained by the beaker, will all of the ice melt? Explain your answer. [3]

energy for a change of state = mass × specific latent heat

specific latent heat of fusion of water = 334 000 J/kg

....

....

....

*12 Gamma radiation can be used to treat cancer tumours.

Explain why gamma radiation can be used to treat cancer. Describe the risks and how the risks are controlled. [6]

....

....

....

....

....

....

....

....

....

....

TOTAL FOR PAPER = 70 MARKS

GCSE
Combined Science

Paper 6: Physics 2

Higher

Time: 1 hour 15 minutes

You may use:

- a calculator
- a ruler.

Instructions

- Use black ink or black ball-point pen. Draw diagrams in pencil.
- Read each question carefully before you start to write your answer.
- Answer **all** questions in the spaces provided.
- Show your working in any calculator question and include units in your answer where appropriate.
- In questions marked with an asterisk (*), marks will be awarded for your ability to structure your answer logically, showing how the points that you make are related or follow on from each other where appropriate.

Information

- The marks for each question are shown in brackets.
  Use this as a guide to how much time to spend on each question.
- The maximum mark for this paper is 70.
- Diagrams are not accurately drawn unless otherwise stated.

Name: ....

1 **(a)** Which of the following is a vector quantity?
Tick (✓) **one** box. [1]

Distance ☐

Mass ☐

Speed ☐

Velocity ☐

(b) A small solid steel cube is taken from the Earth's surface to the International Space Station.

Which of the following properties will change?
Tick (✓) **one** box. [1]

Mass ☐

Surface area ☐

Volume ☐

Weight ☐

2 Look at **Figure 1**, which shows a velocity–time graph of a car travelling on a road.

Figure 1

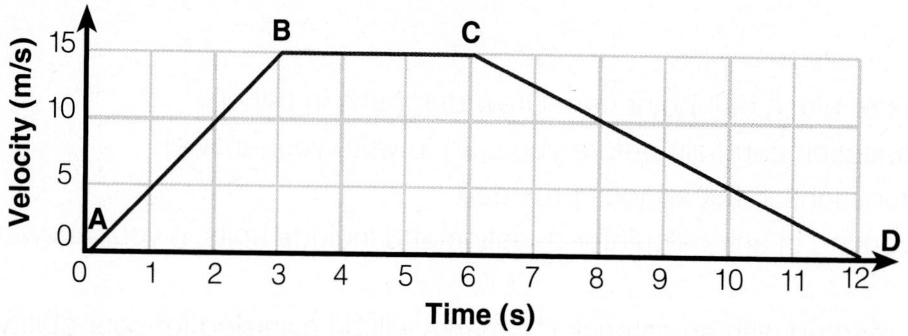

(a) What is the velocity of the car after two seconds? [1]

....

(b) Calculate the acceleration of the car between points **A** and **B**.
Show all your working and include the unit in your answer. [3]

....

....

....

....

(c) Calculate the distance travelled between 0 and 6 seconds. Show all your working. **[3]**

....

....

....

(d) Describe the motion of the car between points **C** and **D**. **[2]**

....

....

3 **(a)** Look at **Figure 2**, which shows a wave.

Figure 2

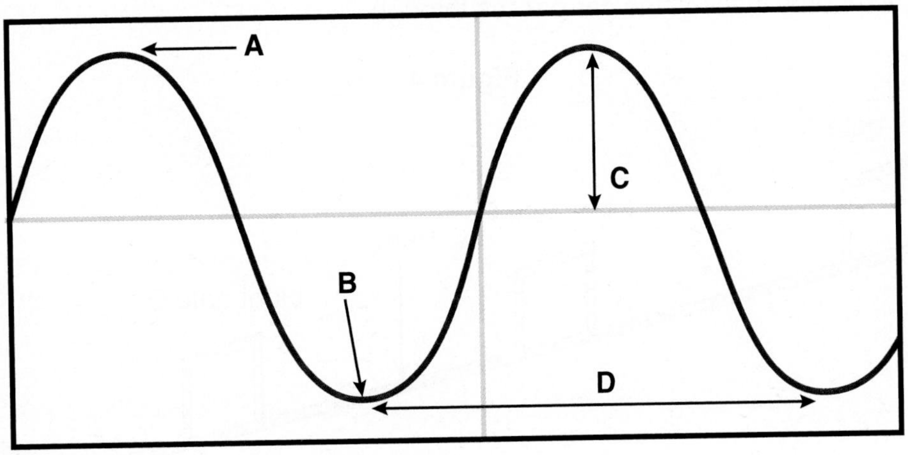

Which letter represents the amplitude of the wave?

Tick (✓) **one** box. **[1]**

A ☐

B ☐

C ☐

D ☐

(b) Microwaves can be used to cook food in a microwave oven, shown in **Figure 3**.
Microwaves are part of the electromagnetic spectrum.

Figure 3

Which other part of the electromagnetic spectrum is commonly used to cook food? **[1]**

....

(c) The microwaves used in ovens have a wavelength of 0.12 m.

The speed of electromagnetic waves is 3×10^8 m/s.

Calculate the frequency of the microwaves used in ovens. Show all your working and include the unit. **[3]**

....

....

....

4 David and Gemma are investigating the motion of a toy car. David says that the car will go down the ramp at constant velocity. Gemma disagrees. She thinks the car will gradually go faster.

They design the experiment shown in **Figure 4** to settle the argument. Light gates are attached to a data logger that records the time the car passes through.

Figure 4

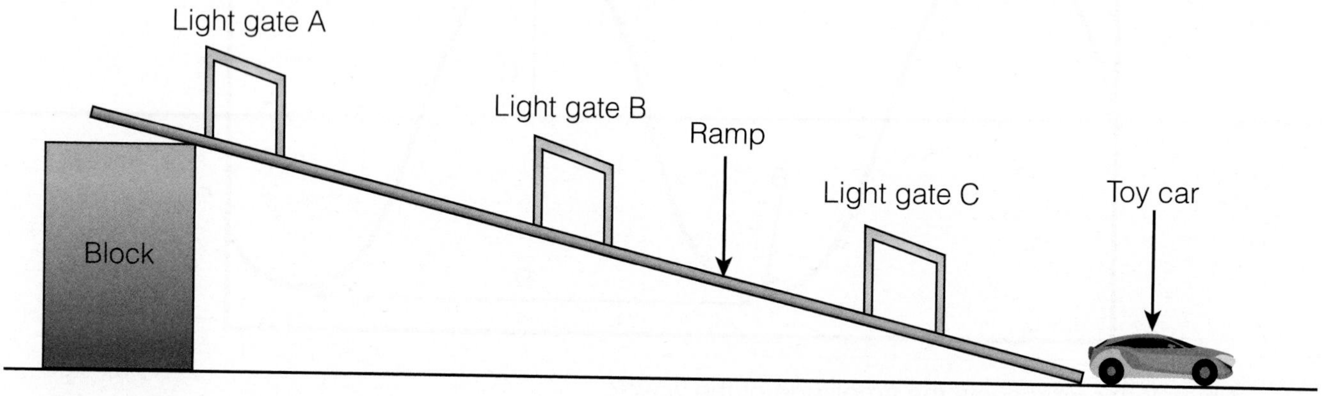

(a) Explain why David and Gemma must use three light gates to time the car. **[1]**

....

....

(b) Describe how they can use the times to decide who is correct. **[2]**

....

....

....

(c) Table 1 shows their results.

Table 1

	Run 1	Run 2	Run 3	Run 4
Time at A (s)	0	0	0	0
Time at B (s)	0.141	0.229	0.152	0.139
Time at C (s)	0.249	0.342	0.270	0.251

Identify the anomalous result.
Tick (✓) **one** box. [1]

Run 1 ☐

Run 2 ☐

Run 3 ☐

Run 4 ☐

(d) The distance between light gate A and light gate C is 1.50 m.

Ignoring the anomalous result, calculate the overall mean speed of the car from the other runs, in m/s.
Show all your working and give your answer to the appropriate number of significant figures. [3]

....

....

....

....

.... m/s

5 **Figure 5** shows an astronaut doing a spacewalk. He can move about using small jets of air blown in directions A, B, C and D. He is stationary.

Figure 5

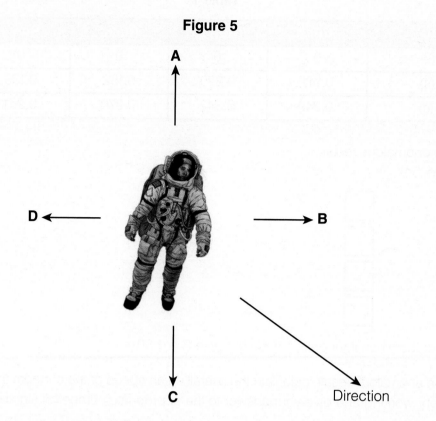

Which two jets should the astronaut use to move in the direction of the arrow shown?
Tick (✓) **one** box.

[1]

A and B ☐

A and D ☐

B and C ☐

C and D ☐

6 **Figure 6** shows a skydiver.

Figure 6

(a) X and Y are forces acting on the skydiver as she falls. Describe what happens to the size of force X as she accelerates. **[1]**

....

(b) The skydiver has a weight of 550 N.

What will be the value of force X when she stops accelerating and falls at terminal velocity? Explain your answer. **[2]**

....

....

(c) **Figure 7** shows the velocity of the skydiver as she descends through the air.

Figure 7

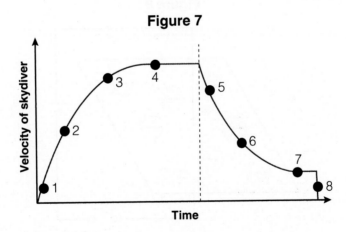

Between which two points does the skydiver open her parachute?
Tick (✓) **one** box. **[1]**

4 and 5 ☐

5 and 6 ☐

6 and 7 ☐

7 and 8 ☐

*(d) Explain why the skydiver has a lower terminal velocity when the parachute is open. **[4]**

....

....

....

....

7 **(a)** **Figure 8** illustrates an experiment to show a magnetic field around a bar magnet using plotting compasses.

Figure 8

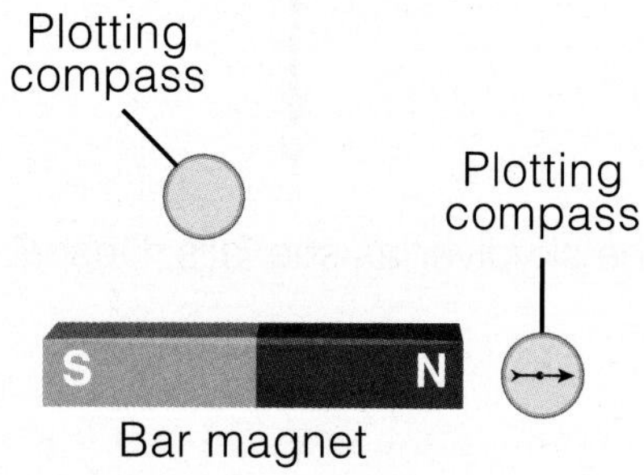

Which direction will the arrow point in the top plotting compass?
Tick (✓) **one** box. **[1]**

(b) Look at **Figure 9**.

Figure 9

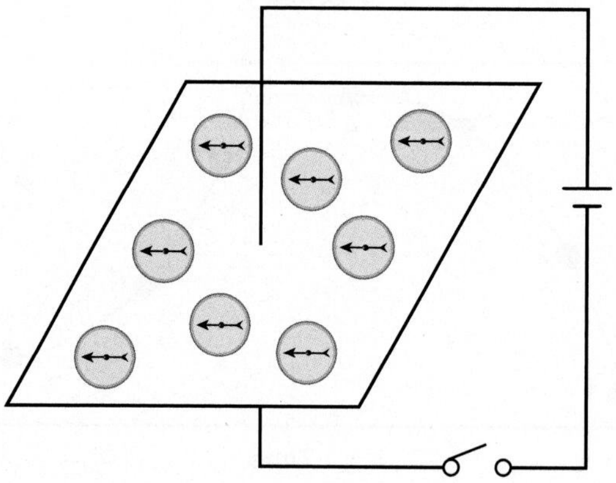

Describe what happens to the plotting compass arrows when the current is switched on. **[2]**

....

....

8 **(a)** Raj is trying to make an electromagnet that will pick up as many paperclips as possible. As shown in **Figure 10**, he adds a soft iron core to the coil.

Figure 10

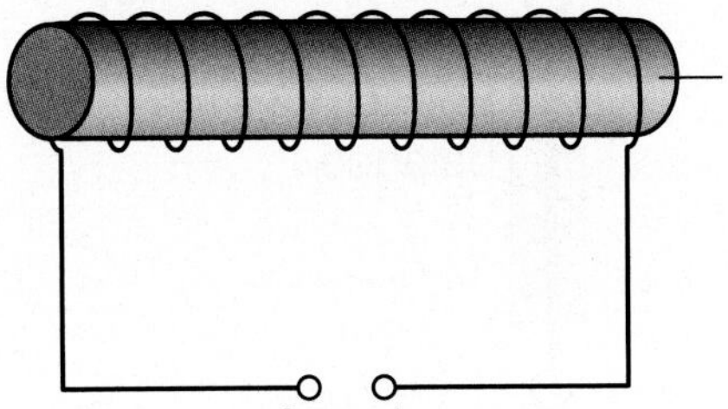

Coil

Iron core

Supply

What **two** other changes could he make to increase the strength of the electromagnet? **[2]**

....

....

(b) As shown in **Figure 11**, a wire is placed in a magnetic field and the current is switched on.

Figure 11

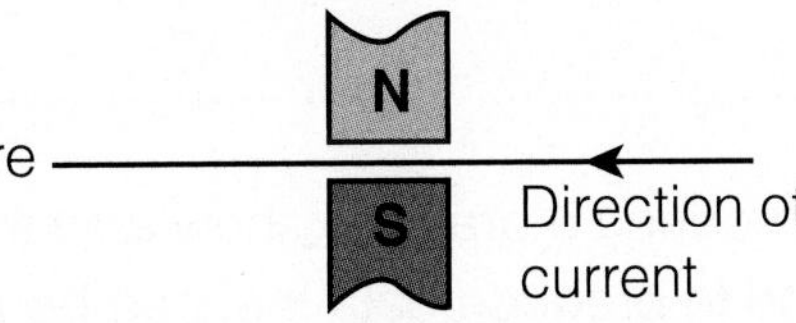

N

Wire

S   Direction of
current

In what direction will the wire move?
Tick (✓) **one** box. **[1]**

Down into the page, away from you   ☐

Towards the north pole of the magnet   ☐

Towards the south pole of the magnet   ☐

Up from the page, towards you   ☐

(c) Explain why the wire only moves once the current is on. **[2]**

....

....

....

9 Navjot and Shaleen are investigating the forces involved in floating and sinking.
Figure 12 shows their experiment.

Figure 12

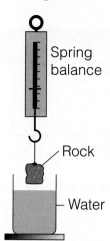

(a) Describe what happens to the reading on the spring balance as Navjot and Shaleen slowly
lower the rock into the water until it is fully submerged. **[2]**

....

....

(b) Navjot and Shaleen want to produce a graph to show how the force changes. Describe how
they can improve their method to produce data that can be plotted on a line graph. **[2]**

....

....

....

(c) Label the axes on **Figure 13** to show how Navjot and Shaleen could plot their results. **[1]**

Figure 13

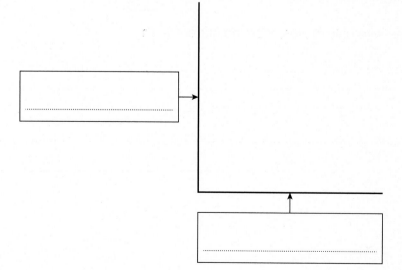

*10 (a) Explain why, theoretically, a ship would rise as it travelled from fresh water into seawater. [3]

....

....

....

....

(b) A diver uses echo sounding from his boat to measure the depth of water, as shown in **Figure 14**.

Figure 14

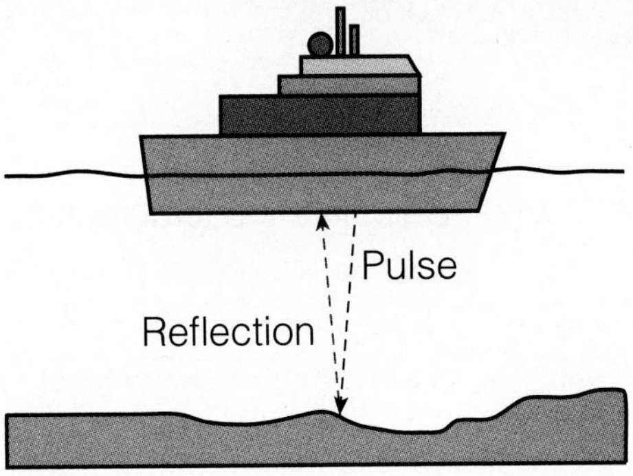

The pulse of sound takes 0.0270 s to be reflected back to be detected on the boat. If sound travels at a speed of 1500 m/s in water, what is the depth of the water in metres under the boat?
Show all your working. [3]

....

....

....

....

(c) Josh wears a pressure gauge on his wrist. He notices that when he dives 10 m under the water, the pressure around him doubles. When he climbs 10 m to the top of the boat the pressure hardly changes at all.

Explain these observations. [2]

pressure = height of column × density of fluid × gravitational field strength

....

....

....

11 **Figure 15** shows a car in a crash test. The car has a mass of 1200 kg.

Figure 15

(a) The car travels at a velocity of 20 m/s. Calculate the total momentum of the car. Show all your working. **[2]**

....

....

....

(b) When the car is crashed into a wall its momentum becomes zero in 0.50 seconds. What is the size of the force that the car exerts on the wall? **[2]**

$$\text{force} = \frac{\text{change in momentum}}{\text{time}}$$

....

....

....

*12 A car driver's awareness of stopping distances is an important part of driving safely.

(a) The thinking distance and the braking distance can be affected by different factors.
Explain how speed, alcohol and road conditions can affect driving safety. **[6]**

....

....

....

....

....

....

....

....

(b) Stopping distance is also affected by tiredness. Why is tiredness not a very good variable for a
scientific experiment? **[1]**

....

13 **Figure 16** demonstrates the motor effect.

Figure 16

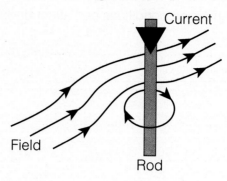

*(a) Use Fleming's left hand rule to explain in which direction the rod will move. **[4]**

....

....

....

....

(b) How could you increase the movement of the wire? **[2]**

....

....

TOTAL FOR PAPER = 70 MARKS

Answers
Biology

CELL BIOLOGY

Page 6

1. **(a)** Nucleus **(1)**
   (b) Mitochondria are too small to see with a light microscope / require an electron microscope / resolution of microscope not high enough. **(1)**
   (c) Liver cells require the release of a lot of energy **(1)**; mitochondria release energy for muscle contraction **(1)**.
   (d) Cell, cytoplasm, nucleus, chromosome, gene **(2 marks for everything in the correct order, 1 mark if cell, cytoplasm and nucleus are in the correct order, reading left to right)**
2. **(a)** Bacteria have a chromosomal loop; have plasmids; no nuclear membrane / nucleus. **(1)**
   (b) Prokaryotic ✓ **(1)**

Page 7

1. **(a)** An undifferentiated cell that can develop into one of many different types. **(1)**
   (b) **Any two from:** nucleus, cytoplasm, cell / plasma membrane, mitochondria, ribosomes. **(2)**
2. **(a)** **Any two from:** therapeutic cloning, treating paralysis, repairing nerve damage, cancer research, grow new organs for transplantation. **(2)**
   (b) **Any one from:** stem cells are sometimes obtained from human embryos and people believe it is wrong to use embryos for this purpose, risk of viral infections. **(1)**
3. $\frac{3600}{30} \times 200$ **(1)** = 24 000 µm **(1)** = 24 mm **(1)**

Page 8

1. **(a)** Line should point to one of the objective lenses / rotating nose cone. **(1)**
   (b) The organelles are too small to see **(1)** / microscope doesn't have a high enough resolving power. **(1)**
   (c) Size of real object = $\frac{3}{400}$ **(1)** = 0.0075 cm **(1)** = 75 µm **(1)**
2. **(a)** Scanning electron microscope **(1)**
   (b) **For 2 marks, accept any answer between 80–100 minutes**; if answer is incorrect, award 1 mark for the idea that there have been 4 divisions.

Page 9

1. **Any two from:** specialised organs carry out a specific job; multicellular organisms are complex and require specialised organs so they can grow larger; single-celled organisms are small enough not to require specialised cells and transport systems. **(2)**

2.

Mitosis	Meiosis
Involved in asexual reproduction	**Involved in sexual reproduction (1)**
Produces clones / no variation (1)	Produces variation
Produces cells with 46 chromosomes	**Produces cells with 23 chromosomes (1)**

3. **(a)** Meiosis **(1)**
   (b) Four cells produced (in second meiotic division) **(1)**
4. **(a)** Benign tumours don't spread from the original site of cancer in the body **(1)**; Malignant tumour cells invade neighbouring tissues / spread to other parts of the body / form secondary tumours. **(1)**
   (b) **Any two from:** not smoking tobacco products; not drinking too much alcohol; avoiding exposure to UV rays; eating a healthy diet; taking moderate exercise / reduce obesity. **(2)**

Page 10

1. $C_6H_{12}O_6 + 6O_2 \longrightarrow 6CO_2 + 6H_2O$ **(1 mark for correct formulae, 1 mark for correct balancing)**
2. **(a)** **Any two from:** larger athletes will use more oxygen due to their higher muscle mass; the adjustment allows rates to be fairly / accurately compared; different athletes may have different masses. **(2)**
   (b) Sprinting has a greater energy demand **(1)**, so more oxygen is needed **(1)**.
   (c) Boris' consumption rate would be lower **(1)** because his lungs, heart and muscles are less efficient at transporting / using oxygen **(1)**.

Page 11

1. **(a)**

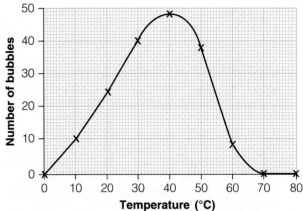

   (2 marks for correct plotting; 1 mark for smooth curve; subtract 1 mark for every incorrect plot)
   (b) The rate of bubbles produced increases until it reaches an optimum / maximum **(1)**; then it decreases rapidly, producing no bubbles at 70°C **(1)**.
   (c) 40°C **(1)**
   (d) **Any two from:** enzyme / active site has changed shape / become denatured; substrate / hydrogen peroxide no longer fits active site, so substrate cannot be broken down; low pH of the stomach **or** amylase works best at alkaline pH and stomach is acid. **(2)**

TRANSPORT SYSTEMS AND PHOTOSYNTHESIS

Page 12

1. **(a)** **Any two named small nutrient molecules**, e.g. glucose, vitamins, minerals, ions, amino acids, oxygen **(2)**.
   (b) **Any one from:** carbon dioxide or urea. **(1)**
2. **(a)** Water moves down concentration gradient from high water concentration to low water concentration **(1)**, across a partially / differentially permeable membrane / plasma membrane **(1)**; gradient maintained by input and output of water at each end of cell line **(1)**.
   (b) Water moving from plant cell to plant cell ✓ **(1)**; A pear losing water in a concentrated solution of sugar ✓ **(1)**; Water moving from blood plasma to body cells ✓ **(1)**.
3. **(a)** Rhubarb cells turgid **(1)**; because water moves into cells due to osmosis **(1)**.
   (b) **Any two from:** plasmolysed; plant cell vacuole extremely small; membrane pulled away from wall. **(2)**

Page 13

1. **(a)** (upper) epidermis **(1)**; **(b)** spongy layer / mesophyll **(1)**;
   (c) air space **(1)**.

2. **Roots:** anchor plant in soil / absorb water and minerals **(1)**
   Stem: supports leaves and flowers, transports substances up and down the plant **(1)**
   Leaf: organs of photosynthesis **(1)**
   Flower: reproductive organs, formation of seeds **(1)**

3. **Any three from:** dead cells without cytoplasm; no end walls; hollow lumen; continuous tubes – all adaptations allow efficient movement of water in columns. **(3)**

4. **(a) Any one from:** phloem have perforated end walls / xylem has no end wall; xylem have hard cell walls (contain lignin) / phloem have soft cell walls. **(1)**
   (b) Aphids extract / eat sugar **(1)**; sugar solution transported in phloem **(1)**.

Page 14

1. **(a)** $\frac{8}{43} \times 100 = 18.6\%$ **(1 mark for calculation, 1 mark for correct answer)**
   (b) B (cold moving air) **(1)**
   (c) Water column in xylem would move upwards / towards the leaves more quickly. **(1)**

2. **This is a model answer, which would score the full 6 marks:** As light intensity increases during the day, the rate of photosynthesis increases in the guard cells. This results in more sugar being manufactured, which raises the solute concentration. Increased potassium ions contribute to increased solute concentration too. The guard cells therefore draw in water from surrounding cells by osmosis, becoming more turgid. This causes the stoma to become wider. The arrangement of cellulose in the cell walls of the guard cells means that there is more expansion in the outer wall, resulting in a wider stoma.

Page 15

1. **(a) (i)** Scotland **(1)**
   (ii) 50 deaths per 100 000 (210 − 160) **(1)**
   (b) Men have higher death rates than women. **(1)**

2. Artery – 3 **(1)**; Capillaries in the body – 4 **(1)**; Vein – 1 **(1)**; Capillaries in the lungs – 2 **(1)**

Page 16

1. **(a)** 924.5 **(2)**; **if the answer is incorrect then showing the working (926 + 923 = 1849, then $\frac{1849}{2}$) will gain 1 mark**
   (b) Rats have different body masses / to standardise results. **(1)**
   (c) As the warfarin dose increases, the time to clot also increases. **(1)**

2. Agree **(1)** because as vital capacity increases the time underwater also increases **(1)**.
   Or disagree **(1)** due to **any one from:** only five subjects / not enough data; need to find divers with higher / lower vital capacities **(1)**.

Page 17

1. **(a)**

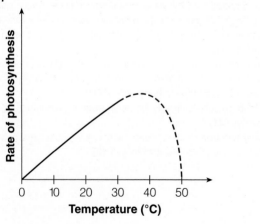

(1)

(b) As the temperature increases, the rate of photosynthesis increases due to more rapid molecular movement and therefore more frequent successful collisions between molecules **(1)**. By 40 °C, the rate peaks and beyond this point enzymes controlling photosynthesis become denatured and the reaction stops **(1)**.

2. **(a)** Leaves **(1)**
   (b) Cellulose: cell walls for support **(1)**; **Protein:** growth / cell membranes / enzyme production **(1)**.
   (c) carbon dioxide + water $\longrightarrow$ glucose + oxygen
   (1 mark for reactants, 1 mark for products)

3. **(a) Any two from:** the increased temperature from the stove will increase photosynthesis rate; increased carbon dioxide concentration will have the same effect; increased photosynthesis means increased starch production / yield. **(2)**
   (b) Any one from: increase light regime, e.g. artificial lighting switched on at night time; increased light intensity / brighter lights. **(1)**

HEALTH, DISEASE AND THE DEVELOPMENT OF MEDICINES
Page 18

1. **(a)** 13.8–14.8% **(1)**
   (b) Glycogen **(1)** found in liver / muscles **(1)**
   (c) Any two from: heart disease / CVD / stroke; cancer; diabetes; asthma / eczema / autoimmune diseases; poor nutrition – named example, e.g. rickets; genetic conditions, e.g. cystic fibrosis; eating disorders, e.g. anorexia; mental health conditions; alcoholism / addiction; named inherited disease. **(2)**

2. **(a)** ×20 **(1)**
   (b) Any one from: low birth weight; premature birth; higher risk of still birth. **(1)**

Page 19

1. **(a)** Pathogen **(1)**
   (b) Cause cell damage **(1)**; the toxins produced damage tissues. **(1)**

2. **(a) Any two from:** diarrhoea; vomiting; dehydration. **(2)**
   (b) Cholera spread by drinking contaminated water **(1)**; water easily contaminated because water supply / sewage systems damaged **or** overcrowding and poor hygiene in disaster zones **(1)**.

3. **(a) Any two from:** malaria is transmitted by the mosquito; warm temperatures are ideal for mosquitos to thrive; stagnant water is an ideal habitat for mosquito eggs to be laid / larvae to survive **(2)**.
   (b) Mosquito: vector **(1)**
   Plasmodium: parasite **(1)**
   (c) Nets will deter mosquitoes / prevent bites / prevent transferral of plasmodium **(1)**; antivirals are ineffective as plasmodium is a protist / not a virus **(1)**.

Page 20

1. **(a)** Antibodies **(1)**
   (b) Pathogens are clumped together to prevent their further reproduction, to make them easier for phagocytes to digest. **(1)**
   (c)

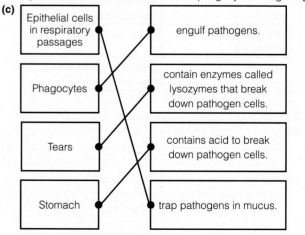

(1 mark for each correct line up to 3 marks, subtract 1 mark for any additional lines)

2. **(a)** **Any one from:** droplet infection / coughing / sneezing / water droplets in breath / aerosol. **(1)**
   (b) Answer in the range 16–17 days **(1)**
   (c) 9 arbitrary units **(1)** (9.5 – 0.5) **(1)**
   (d) Memory cells recognise future invasion of pathogen **(1)**; they can produce the necessary antibodies much quicker, and at higher levels, if the same pathogen is detected again **(1)**.

Page 21
1. **This is a model answer, which would score the full 6 marks:**
   HIV proteins can be triggered and manufactured in existing human cells. The genes that code for the viral proteins are injected into the bloodstream. An adenovirus shell prevents them from being destroyed by the body's general defences. Once inside a cell, the genes instruct it to produce viral proteins that are presented at the cell surface membrane. The body's lymphocytes then recognise these antigens and produce antibodies against them. Memory cells sensitive to the viral proteins are then stored in case the body is exposed to the antigens again.
2. **Accept any three from:** Bacteria are becoming resistant to many modern antibiotics; doctors have in the past over-prescribed antibiotics; mutations in bacteria have led to resistant strains developing; some people don't complete the course of antibiotics. **(3)**

Page 22
1. **(a)** **Any two from:** to ensure that the drug is actually effective / more effective than placebo; to work out the most effective dose / method of application; to comply with legislation. **(2)**
   (b) Double blind trials involve volunteers who are randomly allocated to groups – neither they nor the doctors / scientists know if they have been given the new drug or a placebo **(1)**; this eliminates all bias from the test **(1)**.
   (c) **(i)** Total patients = 226; $\frac{21}{226} \times 100$ **(1)** = 9.29% **(1)**
   (ii) Yes, there are more patients who took the drug and had a cardiovascular event than those who took the placebo. However, there is not a large difference between the groups. **(1)** No, the cardiovascular events could include other conditions apart from heart attacks. **(1)**
   (iii) Rash **(1)**; the difference in numbers of patients who got a rash between the alketronol and placebo groups is quite large **(1)**.

Page 23
1. **(a)** **Mineral deficiency:** lack of nitrates **(1)** **Leaf appearance:** yellow leaves and stunted growth **(1)** (Also accept **Mineral deficiency:** lack of magnesium / potassium / phosphate **(1)**; **Leaf appearance:** chlorosis / discolouration of the leaves **(1)**)
   (b) There is a high concentration of *Chalara* cases in the East of England ✓ **(1)**
   (c) **(i)** **Any two from:** burning trees destroys fungus / *Chalara*; prevents further spores being produced; reduces spread of spores. **(2)**
   (ii) **Any one from:** not all trees removed; trees may produce spores before being detected / destroyed; more spores could be introduced by wind from mainland Europe. **(1)**

COORDINATION AND CONTROL
Page 24
1. homeostasis **(1)**; receptors **(1)**; effectors **(1)**
2. **(a)** Negative feedback **(1)**
   (b) **(i)** It reduces production of ACTH. **(1)**
   (ii) **Any one from:** ineffective nutrient distribution; inability to reduce inflammation; inefficient water control. **(1)**
   (iii) Patient D **(1)**
   (iv) 5 × 7 = 35 µg per litre **(2 marks for correct answer; if answer is incorrect, 1 mark for showing working)**

Page 25
1. **(a)** Nucleus **(1)**; **(b)** Cell body / cytoplasm **(1)**
2. **(a)** Brain **(1)**
   (b) A means of detecting external stimuli, i.e. a **receptor (1)**; transferral of electrical impulses to the CNS, i.e. a **sensory neurone (1)**; a coordinator / control system, i.e. a **brain (1)**.
3. **(a)** Synapse **(1)**
   (b) **Any three from:** transmitter substance released at end of first neurone in response to impulse; travels across synapse by diffusion; transmitter binds with receptor molecules on next neurone; nervous impulse released in second neurone **(3)**.

Page 26
1. **Gland:** pancreas **(1)**   **Hormone:** insulin / glucagon **(1)**
2. **(a)** **Any two from:** after meal, a rise in glucose levels will be detected by device; which will cause hormone implant to release insulin; insulin released to bring blood glucose level down. **(2)**
   (b) People with type 2 diabetes can often control their sugar level by adjusting their diet **(1)**; body's cells often no longer respond to insulin **(1)**.

3.
Gland	Hormones produced
Pituitary gland	**TSH, ADH, LH** and **FSH** (accept any two)
Pancreas	Insulin and glucagon
Thyroid gland	Thyroxine
Adrenal gland	Adrenaline
Ovary	**Oestrogen** and **progesterone**
Testes	Testosterone

(1 mark for each correct line)

Page 27
1. **(a)** 1600 ml **(1)**
   (b) Amounts are equal **(1)**; important that water intake should balance water output to avoid dehydration **(1)**.
   (c) Intake of water greater **(1)**; output from sweating greater **(1)**; water gained from respiration greater **(1)**, as muscles contracting more / respiring more **(1)**.
2. **(a)** B **(1)**
   (b) C **(1)**
   (c) A **(1)**
   (d) Low water levels in blood detected by receptors / in blood vessels / in brain **(1)**; more ADH released by pituitary gland **(1)**; acts on collecting duct / kidney **(1)**; which is stimulated to absorb more water back into bloodstream **(1)**.

Page 28
1. Days 5–14: uterus wall is being repaired **(1)**; egg released at approximately 14 days from ovary **(1)**; days 14–28: uterus lining maintained **(1)**.

2.
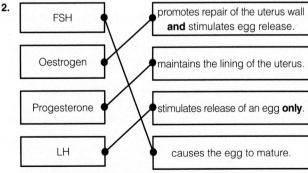

(1 mark for each correct line up to 3 marks)
3. **(a)** Negative feedback **(1)**
   (b) FSH **(1)**
   (c) Progesterone **(1)**
   (d) Continues to be produced **(1)** in large quantities / at high levels **(1)**.

1. **(a)** Leroy and Jane **(1)**
   (b) **Three:** Tim and Margaret, Rohit and Saleema, and Leroy and Jane. **(1)**
   (c) Although irregular ovulation has a lower success rate **(1)**; it affects over twice as many couples (16 × 75 produces a larger total than 7 × 95). **(1)**
   (d) Both methods mean that Jane does not make any genetic contribution **(1)**; egg donation has a high rate of success but can be expensive **(1)**; surrogacy might be cheaper but there is a risk that the surrogate mother might develop an attachment to the baby / want to keep it **(1)**; egg donation requires invasive technique **(1)**.
2. **Any two from:** the contraceptive pill contains hormones that inhibit FSH production; e.g. oestrogen / progesterone; eggs therefore fail to mature; progesterone causes production of sticky cervical mucus that hinders movement of sperm **(2)**.

INHERITANCE, VARIATION AND EVOLUTION
Page 30

1. gametes **(1)**; haploid **(1)**; meiosis **(1)**
2. Meiosis shuffles genes, which makes each gamete unique ✓ **(1)**; Gametes fuse randomly ✓ **(1)**
3. **(a)** **Any three from:** sexual reproduction can be an advantage to a species if the environment changes; asexual reproduction is more advantageous when the environment is not changing; some organisms use both types of reproduction, therefore both have their advantages; sexual reproduction might yield disadvantageous adaptations in an individual when the environment changes. **(3)**
   (b) **Sexual:** male and female parents required; slower than asexual; requires meiosis; more resources (e.g. time, energy) required. **Asexual:** only one parent required; faster than sexual; requires mitosis only; less resources required. **(1)**
   (c) Cytoplasm and organelles duplicated **(1)**; as a 'bud' **(1)**

Page 31

1. The genome of an organism is the entire genetic material present in its adult body cells ✓ **(1)**; The HGP involved collaboration between US and UK geneticists ✓ **(1)**; The project allowed genetic abnormalities to be tracked between generations ✓ **(1)**.
2. **(a)** Organisms with very similar features / chimpanzee and human share equal DNA coding for protein A. **(1)**
   (b) Yeast **(1)**
3. **(a)** **Any two from:** warn women about the risk of cancer ahead of time; enable early and regular screening; enable early treatment; suggest treatment that is targeted. **(2)**
   (b) Other factors may contribute to onset of cancer **(1)**; risk is in terms of a probability (which is not 100%) **(1)**.

Page 32

1. **(a)** T pairs with A; C pairs with G **(both correct for 1 mark)**
   (b) 3 **(1)**
2. **(a)** **Any two from:** UV light; radioactive substances; X-rays; certain chemicals / mutagens. **(2)**
   (b) Base / triplet sequence changed **(1)**; leads to change in amino acid sequence **(1)**; protein no longer has correct shape to perform its job **(1)**.
3. Nucleotides are composed of a sugar-phosphate unit **(1)** and a nitrogenous base **(1)**. The bases are A, T, C and G **(1)**. The sequence of three bases / nucleotides / triplet / codes for a specific amino acid. **(1)**

1. **(a)** 39 **(1)**
   (b) Black is the dominant gene / allele; white is recessive **(1)** **(no marks given for references to 'black chromosome' or 'white chromosome')**; the allele for black fur is passed on / inherited from the father **(1)**.
   (c) Correct genotype or gametes for both parents (Bb and bb) **(1)**; genotype of offspring correct (Bb and bb) **(1)**; correct phenotype of offspring **(1)**.

	b	b
B	Bb Black	Bb Black
b	bb White	bb White

Or

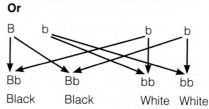

Bb	Bb	bb	bb
Black	Black	White	White

2. **(a)**

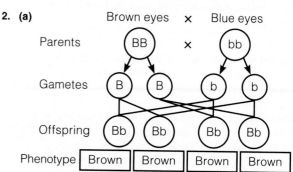

Brown eyes × Blue eyes

Parents	BB	×	bb
Gametes	B B		b b
Offspring	Bb Bb		Bb Bb
Phenotype	Brown	Brown	Brown	Brown

(1 mark will be awarded for the phenotype row and 1 mark for offspring row.)

(b)

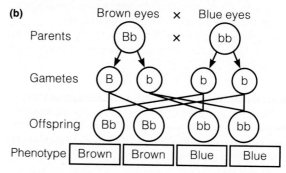

Brown eyes × Blue eyes

Parents	Bb	×	bb
Gametes	B b		b b
Offspring	Bb Bb		bb bb
Phenotype	Brown	Brown	Blue	Blue

(1 mark will be awarded for the phenotype row and 1 mark for the offspring row.)

Page 34

1. (a) 205–215 million years ago **(1)**
   (b) (i) Cretaceous **(1)**
      (ii) They have discovered fossils. **(1)**
   (c) Lizard **(1)**
   (d) Archosaur **(1)**

Page 35

1. (a)

Before Industrial Revolution		After Industrial Revolution	
Pale	Dark	Pale	Dark
1260	107	89	1130

(1 mark for mean numbers before Industrial Revolution; 1 mark for mean numbers after Industrial Revolution)

   (b)

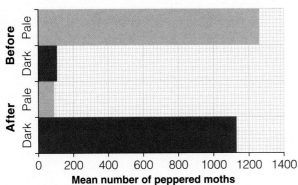

(2 marks for correct plotting of bars; subtract 1 mark for every incorrect plot.)

   (c) The pale-coloured moths could camouflage themselves easily against the silver birch tree bark. **(1)**

   (d) Dark-coloured peppered moths were more camouflaged than pale moths after the Industrial Revolution due to the effects of air pollution **(1)**; dark moths had an increased chance of survival and consequently an increased chance of reproducing and passing on genes **(1)**.

2. Lucy was one of the earliest known hominids to have an upright stance. **(1)**

Page 36

1. (a) **Any three from:** Allow chosen males and females to mate / breed / reproduce together; select offspring from several matings that have high quality wool; allow these sheep to mate together; repeat the process over many generations **(stages must be in sequence)**. **(3)**
   (b) **Any one from:** high quality meat / lamb; thick coat; hardiness / ability to withstand harsh winters; colouration / markings; disease resistance. **(1)**

2. **Any two from:** involves genes, not whole organisms; genes transferred from one organism to another; much more precise in terms of passing on characteristics; rapid production; cheaper than selective breeding; (or reverse argument). **(2)**

3. (a) Crops containing soya can be sprayed with herbicide so weeds are killed rather than soya. **(1)**
   (b) Rice produces carotene, which provides poor populations with vitamin A. **(1)**

4. GM plants may cross-breed with wild plants, resulting in wild plants / weeds that are herbicide-resistant. **(1)**

Page 37

1. (a) **Underline any one of the following**; carnivorous big cats; five toes on their front paws and four toes on their back paws; claws can be drawn back. **(1)**
   (b) Leopards are more closely related to tigers **(1)**; both are the Panthera genus / snow leopards are a different genus **(1)**.

2. Possesses features that are found in reptiles and birds **(1)**; feathers place it with birds but it also has teeth / does not have a beak like reptiles – it is an intermediate form **(1)**.

ECOSYSTEMS

Page 38

1. features; characteristics **(either way round for features or characteristics)**; suited; environment; evolutionary; survival
   (6 words correct = 3 marks, 4 or 5 words correct = 2 marks, 2 or 3 words correct = 1 mark, 1 or 0 words correct = 0 marks)

2. (a)

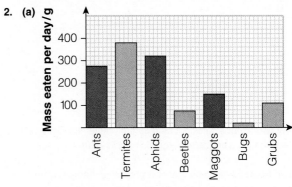

(2 marks for correctly plotting bars, 1 mark for correctly labelling x axis.)

   (b) $\frac{380}{1330} \times 100 = 28.6\%$
   (1 mark for correct answer, 1 mark for showing working)

   (c) It occupies more than one habitat / niche **(1)**; eats a wide variety of food / prey **(1)**.

Page 39

1. (a) Habitat **(1)**
   (b) Ecosystem **(1)**
   (c) **Any one from:** pooter; sweepnet; light trap. **(1)**
   (d) $16 \times 4 \times 5000 = 320\,000$ **(1 mark for correct answer, 1 mark for showing working)**
   (e) (i) Less competition for food between beetles **(1)**; numbers increase as a result **(1)**.
      (ii) Snail numbers would decline. **(1)**

Page 40

1. (a) A producer is an organism that produces its own food. **(1)**
   (b) Sunlight / the Sun **(1)**
   (c) **Any one from:** wasp; ladybird; hoverfly **(1)**

2. (a) There was a slight rise in 1974 **(1)**, but since then the numbers have decreased rapidly **(1)**, and then they have decreased slowly **(1)**. The numbers decreased rapidly between 1976 and 1986 and decreased slowly between 1986 and 2002 **(1)**.
   (b) Farmers have cut down hedgerows and/or trees, so the birds have had nowhere to nest and their food source has been reduced. **(1)**
   (c) **Any one from:** plant more trees; encourage farmers to plant hedgerows; encourage farmers to leave field edges wild as food for birds; use fewer pesticides. **(1)**

Page 41

1. (water) (wind) (wood) **(1)**

2. **Climate zones** shift, causing ecosystems and habitats to change; organisms are displaced and become extinct. **(1)** **Sea levels** rise, causing flooding of coastal regions; islands are inundated and disappear beneath sea level. **(1)** **Ice caps and glaciers** melt and retreat, resulting in loss of habitat. **(1)**

3. (a) See graph **(2)**
(b) Millions (Also accept billions) **(1)**

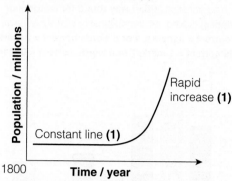

(c) Any two from: insufficient birth control; better life expectancy; better health care; better hygienic practice. **(2)**

Page 42

1. **Any two from:** climate change; new predators; habitat destruction; hunting; competition; pollution. **(2)**
2. **(a)** Cutting down large areas of forest ✓ **(1)**
   (b) Increase in atmospheric carbon dioxide ✓ **(1)**
3. **Any two from:** provide land for agriculture; road building; mining; construction. **(2)**
4. tropical; trees; carbon dioxide; biodiversity; extinct; habitats **(6)**
5. Nutrients in the soil are absorbed by crops and are not replaced. **(1)**

Page 43

1. organisms **(1)**, evaporated **(1)**, precipitation **(1)**
2. **(a)** Tube D **(1)**; because it is warm and moist **(1)**.
   (b) The soil / air / surface of the leaf **(1)**
   (c) So that air / oxygen can get in **(1)**
   (d) Accept one from: they could count the number of whole discs left at the end; they could record what fraction / percentage of leaf discs decayed and find an average; they could measure the percentage decrease in mass of discs by measuring mass before and after time in soil. **(1)**

Page 44

1. **(a)** 160 thousand tonnes (plus or minus 10 000 or in range 150–170) **(1)**
   (b) Overall decrease in numbers **(1)**; temporary rises in 1988–1993 and 2003–2006 **(1)**.
   (c) (i) 1907 tonnes **(1)**
       (ii) Numbers of haddock still declining **(1)**; therefore less fish should be caught in order for fish stocks to recover **(1)**.
   (d) Any one from: increasing mesh size to allow young cod to reach breeding age; increase quotas of other fish species. **(1)**

Chemistry

ATOMIC STRUCTURE AND THE PERIODIC TABLE
Page 45
1. (a) Na **(1)** Cl **(1)**
   (b) Sodium + chlorine $\longrightarrow$ sodium chloride **(reactants 1, products 1)**
   (c) A compound **(1)** as the elements are chemically combined/joined **(1)**
   (d) A mixture **(1)** as the salt and water are together but not chemically combined **(1)**
   (e) Crystallisation **(1)**; simple distillation **(1)**

Page 46
1. (a) The plum pudding model proposed that an atom was a ball of positive charge/today's model has the positive charge contained in the nucleus/protons **(1)** The plum pudding model proposed that electrons were embedded/spread throughout the positive charge/today's model has the electrons in different energy levels/shells surrounding the positive charge **(1)**
   (b) Some of the positively charged particles/alpha particles (when fired at gold foil) were deflected **(1)**
   (c) Niels Bohr suggested that electrons orbit the nucleus at specific distances/are present in energy levels/shells **(1)** His calculations were backed up by experimental results **(1)**
   (d)

Particle	Relative charge	Relative mass
Proton	+1	1
Neutron	0	1
Electron	−1	Negligible/approx. 1/2000

   (4)
   (e) (i) 11 **(1)**
       (ii) 23 **(1)**

Page 47
1. (a) 3 **(1)**
   (b) 2,7 **(1)**
   (c) 7 **(1)** As an atom of fluorine has 7 electrons in its outer shell **(1)**
2. (a) By increasing atomic number **(1)**
   (b) Because they both have similar chemical properties **(1)**
   (c) **Accept two from:** have high melting/boiling points; conduct heat and electricity; react with oxygen to form alkalis; malleable/ductile **(2)**

Page 48
1. (a) The atoms have full outer shells/energy levels **(1)** so they do not bond **(1)**
   (b) It increases **(1)**
2. (a) $2Na_{(s)} + 2H_2O_{(l)} \longrightarrow 2NaOH_{(aq)} + H_{2(g)}$ **(1 for correct formula; 1 for correct balancing in equation. Ignore state symbols, even if wrong)**
   (b) Blue **(1)** as an alkali solution/the hydroxide ion is formed **(1)**
   (c) A potassium atom is larger (than a sodium atom), so the outer electron is further away from the nucleus **(1)** so there is less attraction/the electron is more easily lost **(1)**
3. (a) $2Na_{(s)} + Cl_{(g)} \longrightarrow 2NaCl_{(s)}$ **(1 for each side; any correctly balanced equation scores both marks, e.g. $4Na + 2Br_2 \longrightarrow 4NaBr$)**
   (b) Ionic **(1)**
   (c) chlorine + sodium bromide $\longrightarrow$ sodium chloride **(1)** + bromine **(1)**
   (d) Displacement **(1)** A more reactive element (chlorine) takes the place of a less reactive element (bromine) in a compound **(1)**

STRUCTURE, BONDING AND THE PROPERTIES OF MATTER
Page 49
1. sodium – metallic **(1)**; chlorine – simple molecular **(1)**; sodium chloride – ionic **(1)**
2. (a)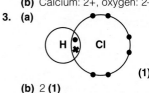
   Ca^{2+} ion [2,8,8] O^{2-} ion [2,8]   **(2)**
   (b) Calcium: 2+, oxygen: 2− **(2)**
3. (a)
   (1)
   (b) 2 **(1)**

Page 50
1. (a) Electrostatic forces **(1)** between the anions and cations
   (b) CaO **(1)** There are an equal number of calcium ions and oxygen ions and so the ratio of each is 1:1 **(1)**
2. (a) A polymer **(1)**
   (b) CH_2 **(1)** There are twice as many hydrogen atoms as carbon atoms, so ratio of carbon to hydrogen is 1:2 **(1)**
3. Giant covalent/macromolecular **(1)** it is a large molecule consisting of atoms that are covalently bonded together with a theoretically infinite structure **(1)**

Page 51
1. (a) 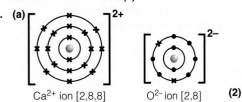 **(1 for regular arrangement of particles, 1 for no gaps between the particles)**
   (b) Intermolecular forces/forces between particles **(1)**
2. (a) It conducts electricity when liquid but not as a solid **(1)**
   (b) The ions **(1)** are not free to move **(1)**
   (c) Giant covalent/macromolecular **(1)** It has a high melting point <u>and</u> does not conduct electricity **(1)**
   (d) There are no free **(1)** charged particles/electrons/ions **(1)**

Page 52
1. (a) The layers **(1)** (of ions) are able to slide over each other **(1)**
   (b) There is a strong attraction **(1)** between the metal cations and the delocalised electrons **(1)**
   (c) The electrons **(1)** are free to move/flow through the structure **(1)**
2. (a) A mixture of metals/a metal mixed with another element **(1)**
   (b) The layers are not able to slide over each other **(1)** because the other atoms are larger and prevent movement **(1)**
3. (a) **Accept two from:** for drug delivery into the body/as lubricants/reinforcing materials, e.g. in tennis rackets **(2)**
   (b) **Accept two from:** tensile strength; electrical conductivity; thermal conductivity **(2)**
   (c) A carbon nanotube is a cylindrical fullerene **(1)**

QUANTITATIVE CHEMISTRY
Page 53
1. (a) Total mass of reactants = total mass of the products, i.e. there is no net mass loss or gain during a chemical reaction **(1)**
   (b)

Substance	A_r / M_r
Al	27
Fe_2O_3	160
Al_2O_3	102
Fe	56

   (4)

2. **(a)** Oxygen **(1)** is added **(1)** to the magnesium
   (b) $2Mg_{(s)} + O_{2(g)} \longrightarrow 2MgO_{(s)}$ **(1 for reactants and products, 1 for correct balancing)**
   (c) The magnesium carbonate loses/gives off **(1)** carbon dioxide **(1)**
   (d) $MgCO_{3(s)} \longrightarrow MgO_{(s)} + CO_{2(g)}$ **(1 for correct formulae products; 1 for balanced equation; ignore state symbols, even if wrong)**

Page 54

1. **(a)** 2.408×10^{24} $(4 \times 6.02 \times 10^{23})$ **(1)**
   (b) 92 g $(4 \times 23$ g$)$ **(1)**
   (c) 0.5 $(11.5 \div 23)$ **(1)**
   (d) 4 moles of Na form 2 moles of Na_2O (2:1 ratio), therefore 0.5 mole of Na forms 0.25 mole of Na_2O mass = 0.5×62 **(1 for correct M_r of Na_2O)** = 31 g **(1)**
2. **(a)** $6 \div 12 = 0.5$, $1 \div 1 = 1$ **(1)**
   $0.5 \div 0.5 = 1$, $1 \div 0.5 = 2$ **(1)**
   Empirical formula = CH_2 **(1)**
   (b) (relative formula mass of empirical formula = 14), $98 \div 14 = 7$ **(1)**
   Molecular formula = C_7H_{14} **(1)**

Page 55

1. **(a)** $(5 \div 200) \times 1000 = 25$ g/dm³ **(1)**
   (b) $5 \div 200 \times 14 = 0.35$ g/dm³ **(1)**
   (Also accept $25 \times \left(\dfrac{14}{1000}\right) = 0.35$ g/dm³

2.

Chemical	Pb	O_2	PbO
Mass from question/g	41.4	3.2	44.6
A_r or M_r	207	**32**	223
Moles = $\dfrac{\text{mass}}{M_r}$	$\dfrac{41.4}{207}$ = 0.2	$\dfrac{3.2}{32}$ = 0.1	$\dfrac{44.6}{223}$ = 0.2
÷ smallest	$\dfrac{0.2}{0.1}$ = 2	$\dfrac{0.1}{0.1}$ = 1	$\dfrac{0.2}{0.1}$ = 2

Balanced equation: $2Pb + O_2 \longrightarrow 2PbO$ **(1 for each row in table, 1 for correct balanced equation)**

CHEMICAL AND ENERGY CHANGES

Page 56

1. **(a)** calcium + oxygen $\longrightarrow$ calcium oxide **(1 for reactants, 1 for product)**
   (b) Calcium gains **(1)** oxygen **(1)** (**Also accept** Oxygen **(1)** is added **(1)** or Calcium loses **(1)** electrons **(1)**)
   (c) Potassium, sodium or lithium (or any other metal in group 1 or strontium, barium or radium) **(1)** the metal is more reactive than calcium/is above calcium in the reactivity series **(1)**
   (d) Metal + calcium oxide $\longrightarrow$ metal oxide + calcium, e.g. potassium + calcium oxide $\longrightarrow$ potassium oxide + calcium **(1 for reactants, 1 for products)**
   (e) Calcium oxide **(1)**
2. **(a)** Electrolysis **(1)**
   (b) $2Fe_2O_3 + 3C \longrightarrow 4Fe + 3CO_2$ **(1 for correct formulae, 1 for balanced equation; allow any correct balanced equation, e.g. $4Fe_2O_3 + 6C \longrightarrow 8Fe + 6CO_2$ ignore state symbols)**
   (c) Reduction **(1)** as electrons are gained **(1)** by the Al^{3+} **(1)**
   (d) K **(1)** as it loses electrons **(1)** $K \longrightarrow K^+ + e$ **(1)**

Page 57

1. **(a)** Zinc sulfate **(1)**
   (b) Zinc oxide + hydrochloric acid $\longrightarrow$ zinc chloride **(1)** + water **(1)**
   (c) Carbon dioxide **(1)** CO_2 **(1)**
   (d) (i) calcium (as it loses electrons) **(1)**
   (ii) H^+ **(1; allow $2H^+$ not hydrogen)**
   (e) When a substance loses **(1)** electrons **(1)**
2. **(a)** **Accept three from:** measure out some sulfuric acid (e.g. 25 cm³ in a measuring cylinder); transfer to a beaker and warm the acid; add copper oxide, stir and repeat until no more copper oxide dissolves; filter the mixture; leave the filtrate somewhere warm/heat the filtrate **(3)**

(b) Nitric acid **(1)**
(c) Soluble salt **(1)**

Page 58

1. **(a)** 1–6 **(1)**
   (b) Hydrogen ion **(1)** H^+ **(1)**
   (c) Hydrochloric acid **(1)**
   (d) $H^+_{(aq)} + OH^-_{(aq)} \longrightarrow H_2O_{(l)}$ **(1 for correct formulae, 1 for correct state symbols)**
2. **(a)** A strong acid completely ionises/fully dissociates in water **(1)** a weak acid partially ionises/dissociates in water **(1)**
   (b) $CH_3COOH_{(aq)} \rightleftharpoons CH_3COO^-_{(aq)} + H^+_{(aq)}$ **or** $CH_3COOH_{(aq)} + aq \rightleftharpoons CH_3COO^-_{(aq)} + H^+_{(aq)}$/$CH_3COOH_{(aq)} + H_2O_{(l)} \rightleftharpoons CH_3COO^-_{(aq)} + H_3O^+_{(aq)}$ **(1 for reactants, 1 for products; ignore state symbols)**
   (c) Lower **(1)** Nitric acid is a stronger acid and so there will be more H^+ ions than in a weak acid and therefore will be more acidic **(1)**
   (d) It increases **(1)** by 1 **(1)**

Page 59

1. **(a)** Cations **(1)**
   (b) Anions **(1)**
   (c) So that the ions **(1)** are free to move **(1)**
   (d) Oxidation **(1)** as electrons are lost **(1)**

2. **(a)**

Solution	Product at anode	Product at cathode
NaCl	H_2	Cl_2
KNO_3	**H_2**	**O_2**
$CuSO_4$	**Cu**	**O_2**
Water diluted with sulfuric acid	**H_2**	O_2

(6)

(b) Hydrogen is produced because hydrogen ions are present in the solution **(1)** and hydrogen is less reactive than sodium **(1)**
(c) $4OH^- \longrightarrow O_2 + 2H_2O + 4e^-$ (or $4OH^- - 4e^- \longrightarrow O_2 + 2H_2O$) **(1 for reactants and products, 1 for balanced equation; ignore state symbols)**

Page 60

1. **(a)**

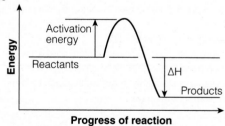

(1 for products lower in energy than the reactants, each correct label scores 1)
(b) More energy is released when the bonds in the product molecules are made **(1)** than is used to break the bonds in the reactant molecules **(1)**
(c) **Accept one from**: any thermal decomposition reaction (e.g. metal carbonate $\longrightarrow$ metal oxide + carbon dioxide); The reaction between citric acid and sodium hydrogencarbonate **(1)**
2. **(a)** ΔH = Bonds broken – bonds formed = 2144 **(1)** – 2348 **(1)** = –204 kJ/mol **(1)**
   (b) Exothermic **(1)** the value of ΔH is negative/more energy is released when the bonds in the product molecules are made than is used to break the bonds in the reactant molecules **(1)**

THE RATE AND EXTENT OF CHEMICAL REACTIONS

Page 61

1. **(a)** 70°C **(1)**
   (b) Increased temperature increases the rate of reaction **(1)** the higher the temperature, the more kinetic energy the particles have and so the frequency of collisions/successful collisions increases **(1)**

2. (a) 1st experiment: 69 ÷ 46 = 1.5 **(1)** cm³/s; 2nd experiment: 18 ÷ 10 = 1.8 **(1)** cm³/s **(1 for correct units)**

(b) The second experiment **(1)** as the rate is greater **(1)**

Page 62

1. (a) The particles must collide with enough energy/the activation energy **(1)**

(b) (i) At a higher concentration, there are more particles per unit volume/in the same volume of solution **(1)** meaning that there will be more collisions **(1)**

(ii) At a higher pressure there are more gas molecules per unit volume **(1)**, which means that there is an increased likelihood of a collision **(1)**

(c)

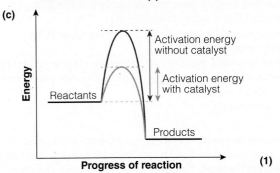

(1)

(d) Catalysts provide an alternative reaction pathway **(1)** of lower activation energy **(1)**

Page 63

1. (a) The blue crystals will turn white **(1)**

(b) hydrated copper(II) sulfate ⇌ anhydrous copper(II) sulfate + water **(allow arrow ⟶ instead of ⇌) (1 for formulae, 1 for balanced equation)**

(c) The white powder will turn blue **(1)**

(d) Exothermic reaction **(1)**

2. (a) A reversible reaction **(1)**

(b) The yield of SO_3 would increase **(1)** higher pressure favours the reaction/shifts the equilibrium **(1)** to the side that produces fewer molecules of gas **(1)**

(c) The yield of SO_3 would decrease **(1)** increasing temperature favours the endothermic reaction **(1)**, which in this case is the reverse reaction **(1)**

ORGANIC CHEMISTRY

Page 64

1. (a) Kerosene **(1)**

(b) It increases **(1)**

(c) 1. Crude oil is heated/boiled **(1)** 2. In the fractionating column there is a temperature gradient (hotter at bottom/cooler at top) **(1)** 3. The hydrocarbons travel up the fractionating column and condense at their boiling point **(1)**

2. (a) C_nH_{2n+2} **(1)**

(b)

```
    H   H   H
    |   |   |
H — C — C — C — H
    |   |   |
    H   H   H
```
(1)

(c) C_4H_{10} **(1)**

(d) Any one from: differ by CH_2 in their molecular formula from neighbouring compounds; show a gradual trend in physical properties; have similar chemical properties **(1)**

Page 65

1. (a) carbon dioxide **(1)** and water **(1)**

(b) $C_3H_8 + 5O_2 \longrightarrow 3CO_2 + 4H_2O$ **(1 mark for correct formulae and 1 mark for correct balancing)**

(c) propane **(1)**

2. (a) Accept one from: silica, alumina, porcelain **(1)**

(b) by heating **(1)** with steam **(1)**

(c) C_4H_{10} **(1)**

(d) Bromine water **(1)** turns colourless / is decolourised / turns from orange to colourless when mixed with the gas **(1)**

CHEMICAL ANALYSIS

Page 66

1. (a) A mixture that has been designed as a useful product **(1)**

(b) No **(1)** something that is chemically pure contains a single element or compound **(1)**

(c) Accept one from: fuels; medicines; foods; fertilisers **(1)**

2. (a) The liquid solvent **(1 allow water)**

(b) red and green **(1)** The spots of these colours are at the same height as spots in ink X **(1)**

(c) $\dfrac{0.5}{4.65} = 0.11$ **(allow 0.10–0.12; 1 mark for working, 1 mark for correct answer)**

(d) It is more accurate/relative amounts of different inks can be determined/smaller quantities can be used **(1)**

Page 67

1. (a) Calcium hydroxide **(1)**

(b) A precipitate **(1)** of $CaCO_3$ **(1)** is formed

2. (a) The litmus paper turns red **(1)** before turning white/being bleached **(1)**

(b) Acidic gas **(1)** as it turns litmus paper red **(1)**

3. (a) Hydrogen gives a squeaky pop **(1)** when exposed to a lit splint **(1)**

(b) Oxygen relights **(1)** a glowing splint **(1)**

(c) $2H_{2(g)} + O_{2(g)} \longrightarrow 2H_2O_{(l)}$ **(1 for formulae of reactants/ products, 1 for correctly balanced equation; ignore state symbols)**

THE EARTH'S ATMOSPHERE AND RESOURCES

Page 68

1. (a) Nitrogen or oxygen **(1)**

(b) The water vapour originally in the atmosphere cooled **(1)** and condensed **(1)** forming the oceans

(c) Green plants/algae **(1)** form oxygen as a waste product of photosynthesis **(1)**

(d) Accept two from: green plants; algae use carbon dioxide for photosynthesis; carbon dioxide is used to form sedimentary rocks; carbon dioxide is captured in oil; coal **(2)**

Page 69

1. (a) Greenhouse gases allow short wavelength radiation from the Sun to pass through the atmosphere **(1)** but absorb long wavelength radiation reflected back from the earth trapping heat **(1)**

(b) Accept one from: methane; water vapour **(1)**

(c) Accept two from: combustion of fossil fuels; deforestation; increased animal farming; decomposing rubbish in landfill sites **(2)**

(d) There are many different factors contributing to climate change **(1)** and it is not easy to predict the impact of each one **(1)**

(e) Accept two from: rising sea levels, which may cause flooding and coastal erosion; more frequent and/or severe storms; changes to the amount, timing and distribution of rainfall; temperature and water stress for humans and wildlife; changes to the food-producing capacity of some regions/changes to the distribution of wildlife species **(2)**

(f) Accept two from: use of alternative energy supplies; increased use of renewable energy; energy conservation; carbon capture and storage techniques; carbon taxes and licences; carbon offsetting/carbon neutrality **(2)**

(g) Accept two from: disagreement over the causes and consequences of global climate change; lack of public information and education; lifestyle changes, e.g. greater use of cars and aeroplanes; economic considerations, i.e. the financial costs of reducing the carbon footprint; incomplete international co-operation **(2)**

Page 70

1. **(a)** Fuel formed in the ground over millions of years from the remains of dead plants and animals **(1)**

(b) Due to incomplete combustion/burning of fuels in a poor supply of oxygen **(1)**

(c) **Accept one from:** forms acid rain; can cause respiratory problems **(1)**

(d) Global dimming **(1)** health problems due to lung damage **(1)**

(e) It combines with haemoglobin/red blood cells **(1)** preventing the transport of oxygen **(1)**

(f) When sulfur burns it react with oxygen to form sulfur dioxide **(1)** Sulfur dioxide dissolves in rain water **(1)** The rain water is now acidic/acid rain which can damage buildings and destroy wildlife **(1)**

Page 71

1. **(a)** The needs of the current generation are met without compromising the potential of future generations to meet their own needs **(1)** Improving agricultural practices/using chemical processes to make new materials **(1)**

(b) Water that is safe to drink **(1)**

(c) Filtration/passed through filter beds to remove solid impurities **(1)** sterilised to kill microbes **(1)**

(d) **Accept three from:** screening and grit removal; sedimentation to produce sewage sludge and effluent; anaerobic digestion of sewage sludge; aerobic biological treatment of effluent **(3)**

2. **(a)**

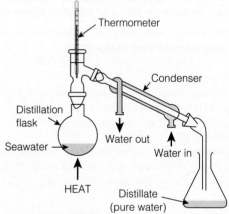

(3 for labelled diagram: seawater in flask, condenser, pure water being collected)

The water from the seawater boils/evaporates at 100°C and is condensed back into water in the condenser **(1)** the salt remains behind in the original flask **(1)**

(b) Reverse osmosis **(1)**

Page 72

1. **(a)** A naturally occurring mineral **(1)** from which it is economically viable to extract a metal **(1)**

(b) We still require the raw materials **(1)** but it is becoming increasingly difficult/uneconomic to mine them in traditional ways **(1)**

(c) Electrical wiring **(1)** as it is a good conductor of electricity **(1)** water pipes **(1)** as it does not react with water/corrode **(1)**

(d) Plants grow in a medium that enables them to absorb metal compounds **(1)** the plants are then harvested/burned leaving ash that is rich in the metal compounds **(1)** the ash is then further treated to extract the metal **(1)**

(e) Bacteria extract metals from low-grade ores producing a solution rich in metal compounds **(1)**

(f) iron + copper sulfate $\longrightarrow$ iron sulfate + copper **(1 for reactants, 1 for products)**

(g) Platinum is less reactive than copper/too low in reactivity/in the reactivity series **(1)** and so is unable to displace copper from copper compounds

(h) Electrolysis **(1)**

Page 73

1. **(a)** **Accept two from:** extracting and processing raw materials; manufacturing and packaging; disposal at end of useful life **(2)**

(b) **Accept two from:** how much energy is needed; how much water is used; what resources are required; how much waste is produced; how much pollution is produced **(2)**

(c) Allocating numerical values to pollutant effects is not always easy or straightforward **(1)** so value judgments have to be made which may not always be objective **(1)**

(d) To support claims for advertising purposes **(1)**

2. Over the life of the product, plastic bags when compared with paper bags (**accept two from the following**): use less energy; use less fossil fuel; produce fewer CO_2 emissions; produce less waste; use less freshwater **(2)** meaning that plastic/polythene bags are less damaging to the environment/have less of an environmental impact **(1)** than paper bags (**Allow reverse argument**)

Physics

FORCES

Page 74

1. **(a)** weight = mass × gravitational field strength
   78 × 10 = 780 N **(1)**
   (b) Mass would be 78 kg (the same as on Earth) **(1)**.
   weight = mass × gravitational field strength
   78 × 1.6 = 124.8 N **(1)**
   (c) The gravitational attraction of Neil is very small **(1)**. The masses of the Moon and Earth are much larger than Neil so the gravitational attraction has no significant effect on them **(1)**.

2. **(a)** 120 − 30 **(1)** = 90 N **(1)**
   (b) (i) 0 N **(1)**
   (ii) Louise is moving at a constant speed **(1)**. As the resultant force is zero she is neither accelerating nor decelerating **(1)**.
   (c) $\dfrac{\text{work done}}{\text{force}} = \dfrac{\text{distance moved along the line of action of the force}}{}$
   $= \dfrac{25\,000}{90}$ **(1)** = 278 m **(1)**

Page 75

1. **(a)** Linear **(1)**
   (b) force = spring constant × extension
   spring constant $= \dfrac{\text{force}}{\text{extension}}$ **(1)**
   $\dfrac{4}{0.01}$ – substitute numbers from graph **(1)**
   = 400 N/m **(1)**
   (c) (i) Non-linear
   (ii) The mass is too large **(1)** so the spring has exceeded the limit of proportionality **(1)**. The extension is no longer proportional to the force applied **(1)**.

Page 76

1. **(a)** Speed only has a magnitude **(1)**. Velocity has a magnitude and a direction **(1)**.
   (b) The direction of the car is changing **(1)**. A change in direction causes a change in velocity but not in speed **(1)**.
   (c) 1.5 m/s – Walking
   6 m/s – Bicycle
   20 m/s – Car
   250 m/s – Plane **(4 marks if fully correct; 1 mark per match)**

2. **(a)** speed $= \dfrac{\text{distance}}{\text{time}}$ **(1)** $= \dfrac{180}{13} = 14$ m/s (2 s.f.) **(1)**
   (b) speed $= \dfrac{\text{distance}}{\text{time}}$ **(1)**
   time $= \dfrac{\text{distance}}{\text{speed}}$
   $= \dfrac{500}{14} = 36$ seconds (2 s.f.) **(1)**

Page 77

1. **(a)** speed $= \dfrac{\text{distance}}{\text{time}}$ **(1)** $= \dfrac{3500}{350} = 10$ m/s **(1)**
   (b) gradient $= \dfrac{1750}{50 \text{ seconds}}$ **(1)** $= 35$ m/s **(1)**

2. **(a) (i)** 250 – 300 seconds **(1)**
   (ii) gradient $= \dfrac{100}{50}$ **(1)** $= 2$ m/s² **(1)**
   (b) area under graph $= \dfrac{(36 \times 50)}{2}$ **(2)**
   $= 900$ m **(1)**

3. A skydiver accelerates due to their weight **(1)** exceeding the air resistance **(1)**. As they reach terminal velocity **(1)** their weight balances the air resistance, or the resultant force is zero, so they stop accelerating and maintain a constant speed **(1)**.

Page 78

1. **(a) (i)** A stationary object will begin moving and accelerate **(1)**.
   (ii) A moving object will either accelerate or decelerate **(1)**.
   (b) Newton's first law **(1)**
   (c) Inertia **(1)**

2. **(a)** The acceleration of an object is proportional to the resultant force **(1)**.
   (b) (i) resultant force = mass × acceleration **(1)**
   $= 110 \times 5$
   $= 550$ N **(1)**
   (ii) The resultant forces are unbalanced **(1)**. The force opposing the cyclist's movement is greater than the force of the cyclist's movement **(1)**. When she stops the resultant force on her is zero **(1)**.

3. Whenever two objects interact, the forces they exert on each other are equal and opposite **(1)**.

4. A vehicle with a mass over the limit would have a weight downwards greater **(1)** than the force upwards from the bridge **(1)**. There is a resultant force downwards and the bridge breaks **(1)**.

Page 79

1. **(a) (i)** Thinking distance: At higher speeds the car travels further in the time it takes to react or process the information and apply the brakes **(1)**.
   (ii) Braking distance: The car travels further in the time it takes the brakes to stop the car **(1)**.
   (b) **Two from**: tiredness; distraction; drugs; alcohol **(2)**
   (c) (i) B, D, C, A **(1)**
   (ii) Test B was done at a higher speed than A so the car had greater kinetic energy **(1)**. This meant the work done was also much greater in order to reduce this kinetic energy **(1)**. This meant there was much greater friction between the brakes and the wheel and so there was a much higher temperature in B than A **(1)**.

2. **(a)** momentum = mass × velocity **(1)**
   $= 50 \times 20$
   $= 1000$ kg m/s **(1)**
   (b) The momentum will decrease. **(1)**

ENERGY

Page 80

1. **(a)** g.p.e. = mass × gravitational field strength × height **(1)**
   $= 60 \times 10 \times 3900$
   $= 2\,340\,000$ J or 2340 kJ **(1)**
   (b) kinetic energy = 0.5 × mass × (speed)² **(1)**
   $= 0.5 \times 60 \times 55^2$
   $= 90\,750$ J / 90.75 kJ **(1)**
   (c) kinetic energy = 0.5 × mass × (speed)² **(1)**
   $= 0.5 \times 60 \times 5^2$
   $= 750$ J **(1)**
   (d) 0 J **(1)**

2. **(a)** Copper **(1)** as it has the lowest specific heat capacity. **(1)**
   (b) Temperature change = 35 − 16 = 19°C **(1)**
   $\dfrac{(\text{change in thermal energy} \div \text{temperature change})}{\text{specific heat capacity}} = \text{mass}$
   $= \dfrac{(76\,000 \div 19)}{390}$ **(1)**
   $= 10$ kg (2 s.f.) **(1)**

Page 81

1. transferred ✓ **(1)** dissipated ✓ **(1)**

2. **(a)** House B **(1)** has the lowest heating cost **(1)**. It is well insulated so loses less heat through the walls **(1)**.
   (b) No **(1)** not all heat is lost through walls **(1)**. Insulation in other areas could affect heating cost as could the efficiency of the boiler providing the heat **(1)**.
   (c) efficiency $= \dfrac{\text{useful output energy transfer}}{\text{useful input energy transfer}}$ **(1)**
   useful output energy transfer = efficiency × useful input energy transfer
   $= 0.65 \times 5400 = 3510$ kJ **(1)**

Page 82

1. Renewable energy: **Three from**: solar; wind; wave; tidal; geothermal; hydroelectric
   Non-renewable energy: **Three from**: coal; oil; gas; nuclear
   (3 marks if fully correct; 2 marks if five correct; 1 mark if three correct)
2. (a) Over time the percentage of electricity generated by wind turbines has increased **(1)**.
   (b) Renewable energy so will not run out like fossil fuels **(1)**, doesn't release carbon dioxide or other pollutants **(1)**.
   (c) No **(1)**. Wind doesn't blow all the time **(1)** so we will always need another source of energy to provide 'back up' and supply electricity when wind turbines aren't turning **(1)**.
   (d) Some people think they are ugly and spoil views of landscapes **(1)**.

WAVES
Page 83

1. (a) A longitudinal wave **(1)**
   (b) wave speed = frequency × wavelength **(1)**
   wavelength = $\dfrac{\text{wave speed}}{\text{frequency}} = \dfrac{1482}{120} = 12.4$ metres **(1)**
   (c) Water is denser than air **(1)** so the sound wave is propagated faster **(1)**.
   (d) Water waves are transverse waves where the oscillations are perpendicular to the direction of energy transfer **(1)**. Sound waves are longitudinal waves where the oscillations are parallel to the direction of energy transfer **(1)**.
2. Amplitude – The maximum displacement of a point on a wave away from its undisturbed position. **(1)**
   Wavelength – The distance from a point on one wave to the equivalent point on the adjacent wave. **(1)**
   Frequency – The number of waves passing a point each second. **(1)**

Page 84

1. A: Radio **(1)** B: Infrared **(1)** C: Gamma **(1)**
2. (a) A: Angle of incidence **(1)**, B: Angle of refraction **(1)**
   (b) The light changes speed as it moves between media **(1)** as water is optically denser than air **(1)**. The wave travels slower in an optically denser medium so bends towards the normal **(1)**.
   (c) The light would bend away from the normal **(1)**.

Page 85

1. Radio waves – Television – Don't require a direct line of sight between transmitter and receiver.
   Ultraviolet – Energy efficient lamps – Require less energy than conventional lights.
   Infrared – Heating a room – Thermal radiation heats up objects.
   (3 marks if fully correct; 1 mark for each correct match of type, use and why suitable)
2. (a) Changes in the nucleus of a radioactive atom **(1)**.
   (b) Sterilisation of equipment or other suitable answer **(1)**.
   (c) They can cause gene mutations **(1)** and cancer **(1)**.
3. Radiation dose **(1)**
4. (a) X-rays **(1)**
   (b) X-rays are absorbed differently by different parts of the body **(1)**. More are absorbed by hard tissues, e.g. bone **(1)** and less are absorbed by soft tissues **(1)**.

ELECTRICITY
Page 86

1. (a) fuse **(1)**
   (b) thermistor **(1)**
   (c) switch (closed) **(1)**
2. (a) X: Filament lamp **(1)**, Y: LED **(1)**, Z: Resistor **(1)**
   (b) 5A **(1)**
   (c) The flow of electrical charge **(1)**
   (d) charge flow = current × time **(1)**
   time = $\dfrac{\text{charge flow}}{\text{current}} = \dfrac{780}{5}$
   = 156 seconds **(1)**

Page 87

1. (a) Ohmic conductor **(1)**
   (b) Temperature **(1)**
   (c)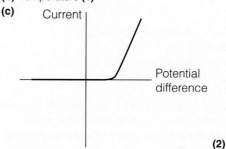

   (2)
2. $\dfrac{\text{potential difference}}{\text{resistance}}$ = current **(1)** = $\dfrac{15}{4}$
   = 3.75 A **(1)**
3. The potential difference across a component and the current through it are required to calculate its resistance **(1)**. The ammeter is wired in parallel not in series so can't be used to measure the current **(1)**. The voltmeter is wired in series not in parallel so can't be used to measure the potential difference **(1)**.

Page 88

1. (a) A series circuit **(1)**
   (b) Total resistance = 4 + 5 + 2 **(1)**
   = 11 Ω **(1)**
2. (a) 2A **(1)**
   (b) 8V **(1)**
   (c) resistance = $\dfrac{\text{potential difference}}{\text{current}} = \dfrac{8}{2}$
   = 4 Ω **(1)**
   (d) The total resistance of all the components would be greater **(1)**. The total resistance of components wired in series is the sum of the individual resistors **(1)**. In parallel the total resistance is always lower than the resistance of an individual resistor **(1)**.

Page 89

1. (a) Live wire – Brown – Carries the alternating potential difference from the supply.
   Neutral wire – Blue – Completes the circuit.
   Earth wire – Green and yellow stripes – Only carries a current if there is a fault.
   (3 marks if fully correct; 1 mark for each correct match of wire, colour and function)
   (b) It is impossible for the case of the appliance to become live as it is plastic **(1)** and it is impossible for the live wire to come into contact with the case **(1)**.
2. (a) Alternating current **(1)**. The sander is plugged into the mains electricity which is a.c. **(1)**.
   (b) The circuit breaker can be easily reset **(1)**. The circuit breaker operates much faster than a fuse **(1)**.

Page 90

1. (a) power = current² × resistance **(1)** = $3^2 × 76.7$
   = 690.3 W **(1)**
   (b) Electrical **(1)** to kinetic and heat **(1)**.
   (c) energy transferred = power × time **(1)**
   time = $\dfrac{\text{energy transferred}}{\text{power}} = \dfrac{180\,000}{690.3} = 261$ seconds **(1)**
   (d) The potential difference is high in cables to reduce heat loss **(1)** and improve efficiency **(1)**. The potential difference is reduced in homes for safety **(1)**.

MAGNETISM AND ELECTROMAGNETISM
Page 91

1. (a) S **(1)**
   (b) N **(1)**
   (c) N **(1)**
2. (a) Magnetic field **(1)**
   (b) Attraction **(1)**
   (c) Force would be stronger at A than at B **(1)**. A is closer to the poles of the magnet / the magnetic field lines are closer together at A **(1)**.

3. The Earth's core is magnetic **(1)** and produces a magnetic field **(1)** so attracts the bar magnet **(1)**.

Page 92

1. The magnetic field around a solenoid is the same shape as the magnetic field around a bar magnet. ✓ **(1)**
   A solenoid with an iron core is an electromagnet. ✓ **(1)**
2. **(a)** The direction of the current **(1)** and the direction of the magnetic field **(1)**.
   (b) The motor effect **(1)**
   (c) force = magnetic flux density × current × length **(1)**
   $$\text{magnetic flux density} = \frac{(\text{force} \div \text{current})}{\text{length}} = \frac{(25 \div 15)}{0.3}$$
   $$= 5.6 \text{ T } \textbf{(1)}$$
3. In the fan there is a coil of wire carrying an electrical current **(1)** in a magnetic field **(1)**. The coil of wire experiences a force that causes the coil to rotate, so it now has kinetic energy **(1)**.

PARTICLE MODEL OF MATTER

Page 93

1. **(a)** $\text{density} = \dfrac{\text{mass}}{\text{volume}}$ **(1)**
   $$\text{density} = \frac{5}{0.002}$$
   $$= 2500 \text{ kg/m}^3 \textbf{(1)}$$
   (b) The density would decrease **(1)** as gas was released **(1)**. The mass of the fire extinguisher would decrease but the volume would remain the same **(1)**.
2. **(a)** Differences in temperature would affect the pressure of the gas **(1)**.
   (b) Differences in volume would affect the pressure **(1)**.
3. **(a)** 3 kg of water **(1)**. Mass is conserved during changes of state **(1)**.
   (b) The change has been reversed and the ice has regained its original properties **(1)**. This occurs in physical changes but not in chemical changes **(1)**.

Page 94

1. **(a) (i)** 10°C **(1)**   **(ii)** 60°C **(1)**
   (b) The temperature is staying the same **(1)** so it is changing state **(1)** and the stored internal energy is changing **(1)**.
   (c) energy for a change of state = mass × specific latent heat **(1)**
   $$= 0.5 \times 250\,000$$
   $$= 125\,000 \text{ J } \textbf{(1)} \text{ or } 125 \text{ KJ}$$
2. **(a)** 329K **(1)**
   (b) 0 Kelvin is –273°C, –300°C is below this **(1)**. 0 Kelvin is absolute zero and at this temperature the particles in the object have no kinetic energy so aren't moving, so it's impossible to have a temperature below this. **(1)**

ATOMIC STRUCTURE

Page 95

1. **(a) (i)** 6 **(1)**
        (ii) 6 **(1)**
        (iii) 6 **(1)**
   (b) (i) Isotopes **(1)**
        (ii) They have different numbers of neutrons **(1)**.
2. **(a)** Nuclear model has a nucleus with electrons orbiting **(1)**. Plum pudding model suggested atom was a ball of positive charge with negative electrons embedded or scattered in it **(1)**.
   (b) It provided evidence of existence of neutrons **(1)**.

Page 96

1. **(a) (i)** A: Alpha **(1)**, B: Gamma **(1)**, C: Beta **(1)**
        (ii) Most to least: Alpha, Beta, Gamma **(2 marks if fully correct; 1 mark for two correct)**
2. **(a)** Alpha **(1)**
   (b) A: 92, B: 234, C: 4, D: 2 **(3 marks if fully correct; 2 marks for three correct; 1 mark for two correct)**
   (c) There is no effect on the mass or the charge of the nucleus **(1)**.

Page 97

1. contamination; decay; emit; Irradiation; radiation; radioactive **(3 marks if fully correct; 2 marks if one error; 1 mark if two errors)**
2. **(a) (i)** 20 days **(1)**
        (ii) This is the length of time it takes for the activity (counts per second) to halve **(1)**.
        (iii) Counts per second at 20 days = 250
        $$\text{Counts per second at 40 days} = \frac{250}{2} = 125 \textbf{ (1)}$$
        $$\text{Counts per second at 60 days} = \frac{125}{2} = 62.5 \textbf{ (1)}$$

Paper 1: Biology 1

Question number		Answer	Notes	Marks
1	(a)	Lactic acid		1
	(b)	Carbon dioxide		1
2	(a)	Her muscles can release more energy.		1
	(b)	**Cause:** build-up of lactic acid (in muscles) **Recovery:** heavy breathing / panting over a period of time		1 1 1
	(c)	**Any three from:** More blood to the muscles. More oxygen supplied to muscles. Increased energy transfer in muscles. More lactic acid to be removed from muscles.		3
3		cell, tissue, organ, system		1
4		White blood cells		1
5	(a)	Right ventricle		1
	(b)	Blood is pumped at a higher pressure. / Blood is pumped a greater distance.		1
	(c)	They stop the backflow of blood.	Allow: they stop blood flowing in the wrong direction.	1
	(d)	$\dfrac{22\,100}{105\,650} \times 100 = 20.91812\%$ 21% to two significant figures		1 1
	(e)	**Any two from:** Low(er) fat intake Low(er) salt intake Regular exercise Healthy body weight Low stress levels		2
6	(a)	It's a physical barrier – prevents pathogens from entering. It produces antimicrobial secretions (peptides) to kill microorganisms.		1 1
	(b)	640 	Time	Bacteria
---	---
0	10
20	20
40	40
60	80
80	160
100	320
120	640		 Allow 1 mark for: number doubles six times in two hours.	1 1
	(c)	Measles is caused by a virus. Antibiotics only treat bacterial infections.		1 1
	(d)	$\dfrac{10}{5000} = 0.002 \times 1000$ $= 2\ \mu m$		1 1
	(e)	**Any two from:** Boil / sterilise all water. Isolate infected individuals. Any basic hygiene measure involving water.	Allow: wash hands / toilets away from water supply, etc.	2
	(f)	There is less chance of contracting the disease as it reduces possible contact with infected people. The disease cannot spread as people are immune.		1 1
7		Cell wall		1

Question number	Answer	Notes	Marks
8	**Any four from:** Sterilise wire / inoculating loop in flame. Transfer sample of pond water to agar. Seal lid with a cross of tape. Turn upside down (to stop condensation). Incubate at 25 °C.		4
9 (a)	Moral / religious / ethical reasons (e.g. use of embryos to obtain stem cells, it interferes with natural processes, 'plays God' or is against religious beliefs) Possibility of virus transfer		1 1
(b)	Differentiation		1
10 (a)	**Any two from:** Guard cells fill with water by osmosis. Guard cells change shape. Guard cells become turgid.		2
(b)	The net movement of particles from an area in which there is a high concentration to one of low concentration.	Allow molecules instead of particles. Allow 1 mark for either of the following: The (random / net) movement of particles. Particles spread out / particles mix up.	1 1
11 (a)	**Any two from:** Long Thin Tubes / hollow / no end walls		2
(b)	**Any two from:** Thin cell wall Large surface area Long, hair-like structure		2
12 (a)	To stop evaporation (of the water)		1
(b)	2.5 cm³ per day $\dfrac{50 - 40}{4}$	Allow 1 mark for: total volume lost = 10 cm³	1 1
13	**Any three from:** Starch – for storage Fat / oil – for storage Cellulose – to make cell walls Amino acids – to make protein	To score a mark, both the substance and reason must be noted.	3
14	Water has moved / diffused into sugar solution from a dilute to a more concentrated solution. Visking tubing is partially permeable.		1 1 1
15	Make up 'acid rain' solution (e.g. a dilute sulfuric acid or nitric acid solution, or a solution with a pH in the range 2–6). Minimum of two groups of infected roses – spray one group of roses with acid solution. Control variable, e.g. same place in garden.		1 1 1

Question number	Answer	Notes	Marks
16 (a)	Higher temperature leads to faster growth. More carbon dioxide leads to faster growth.		1 1
(b)	$6H_2O$ $6O_2$		1 1
(c)	**Level 3:** Correct patterns for linking both temperature and distance to *rate* of photosynthesis. Explanations linked to light intensity, energy and enzymes. **Level 2:** One correct pattern / prediction outlined for both distance and temperature. One correct explanation linked to one pattern / prediction. **Level 1:** One correct pattern / prediction outlined for either distance or temperature. One correct explanation linked to the pattern / prediction. No relevant content **Indicative content** **Patterns / predictions** • Fewer bubbles / less photosynthesis with increasing distance • More bubbles at 35 °C • Fewer / no bubbles at 55 °C • Rate of photosynthesis decreases with increasing distance • Rate of photosynthesis increases with temperature up to around 45 °C • Photosynthesis stops at higher temperature **Explanations** • Increase in distance reduces the amount of light • Less light, less photosynthesis • Higher temperature up to around 45 °C – more photosynthesis • More energy increases photosynthesis • Rate of photosynthesis decreases with decrease in light intensity • Correct mention of inverse square law • Rate of photosynthesis increases with increasing energy • Above 45 °C enzymes are denatured and so little or no photosynthesis		5–6 3–4 1–2 0
17	**Any two from:** Viruses are found inside cells. Damaging a virus also damages cells. Viruses have a high mutation rate.		2

Paper 2: Biology 2

Question number	Answer	Notes	Marks
1	A		1
2	Disease		1
3	**Any two from:** Small ears (reduce heat loss) Thick fur (reduces heat loss) Thick fat / blubber (reduces heat loss) Sharp claws (for holding onto prey) Hair on pads of feet (to grip ice)		2
4	$\frac{15 \times 21}{4} = 78.75$ 79 (nearest whole number)		1 1
5 (a)	*Strigops*		1
(b)	Primary consumer		1
6	Less resistant bacteria are killed first / survival of the fittest. If course not completed, resistant bacteria survive. Resistant bacteria then reproduce.		1 1 1
7	Bacteria reproduce very quickly. They form clones / exact copies of the DNA. Bacteria DNA is easily altered / plasmids easily modified.		1 1 1
8	**Any two from:** Smoking Obesity Viruses UV exposure Excessive alcohol intake		2
9	**Any two from:** There was not enough evidence at the time. The mechanism through which variation was passed on was not known. It challenged religious beliefs.		2
10	D		1
11 (a)	Any figure in the range 12.9–15.1. Trend – darker colour to more mass loss. Darker colour – higher temperature / faster decay.		1 1 1
(b)	**Any two from:** Grass from same place Same mass of grass cuttings in each bag Same material for bag Same thickness of material for bag Same thickness / packing of material in the bag		2
12 (a)	Female reaction time better than male / females have faster reaction time than males.		1
(b)	$\frac{0.31 + 0.32 + 0.29 + 0.32}{4}$ Answer is 0.31		1 1
(c)	Rapid / fast Automatic / done without thinking		1 1
(d)	Impulse causes release of chemical / neurotransmitter from ends of sensory neurone. Chemical / neurotransmitter diffuses across gap. Chemical / neurotransmitter binds to receptor molecules on next / relay neurone. This triggers electrical impulse in next / relay neurone.		1 1 1 1

Question number	Answer	Notes	Marks
13 (a)	No The best chance is at or just after 14 days. That's when the egg is released / level of LH (luteinising hormone) is at its highest.	'No' on its own is not enough to obtain 1 mark.	0 1 1
(b)	Maintains lining of uterus Inhibits both FSH and LH		1 1
(c)	**Any two from:** They inhibit FSH. No eggs mature (and they cannot be fertilised). Progesterone causes production of sticky mucus, which hinders movement of sperm.		2
(d)	**Any two from:** Success rates are not high Can lead to multiple births Multiple births have a risk for mother / babies Very stressful		2
14	Select the cows and bulls that have the milk qualities required. Breed them. Select the offspring that have the best milk qualities. Repeat over a number of generations.		1 1 1 1
15 (a)	39		1
(b)	Mitosis **Any two from:** Chromosomes are copied. Two full sets of chromosomes are made. Each set moves to the opposite end of the cell. **Any one from:** Cell divides in two. Each new cell contains a complete set.		1 2 1
16 (a)	Both parents: Ff Four offspring: FF Ff Ff ff	Upper and lower case letters must be distinguishable.	1 1
(b)	25% or 1 in 4		1
(c)	Heterozygous		1
17 (a)	**Level 3:** Correct descriptions of all four control methods. **Level 2:** Correct description of control method 1 **and** one other control method (2, 3 or 4). **Level 1:** One correct description of any control method. No relevant content **Indicative content** **Control method 1** Pancreas monitors glucose level Pancreas produces insulin if glucose level high Glucose converted to glycogen Glycogen stored in liver / cells **Control method 2** Pancreas produces glucagon if glucose level low Glycogen converted to glucose by glucagon **Control method 3** Type 1: blood glucose levels tested regularly Insulin injections if glucose level high **Control method 4** Type 2: diet control Regular exercise	Allow 3 marks for a full description of methods 3 and 4 only.	5–6 3–4 1–2 0

For question 16 (a), the Punnett square:

	F	f
F	FF	Ff
f	Ff	ff

Question number	Answer	Notes	Marks
(b)	**Any two from:** They: filter the blood remove water from the blood remove ions from the blood reabsorb glucose into the blood reabsorb water into the blood reabsorb ions into the blood.		2
(c)	Proteins broken down to amino acids. Excess amino acids form ammonia (in liver). (Toxic) ammonia converted to urea. Urea removed by kidneys / less protein – less urea.		1 1 1 1
18	Burning trees produces more carbon dioxide in the air. / Less photosynthesis means more carbon dioxide in the air. Carbon dioxide is a greenhouse gas / causes global warming.		1 1
19	**Any two from:** Stabilises ecosystems Preserves food chains Preserves possible future crops Preserves possible future medicines Preserves interdependence of species		2

Paper 3: Chemistry 1

Question number	Answer	Notes	Marks
1 (a)	Each row begins with elements with one outer electron.		1
(b)	Their boiling points increase as you go down the group.		1
(c)		Give 1 mark for an X anywhere in the grey area.	1
2 (a)	They make an alkali when reacted with water.		1
(b)	The outer electron in sodium is further from the nucleus.		1
	Less energy needed to remove the outer electron / the outer electron is less tightly held / less attractive force from nucleus to outer electron.		1
(c)	Electrons are delocalised / sea of electrons,		1
	which means they are free to move		1
3 (a)	High melting point and slippery		1
(b)	2.9×10^7 cm^2	This is how the answer is calculated: volume = area × height (thickness) $1 = \text{area} \times 3.4 \times 10^{-8}$ $\text{area} = \dfrac{1}{3.4 \times 10^{-8}}$ $= 2.9 \times 10^7$ cm^2	1
(c)	**Any two from:** high (tensile) strength high electrical conductivity high thermal conductivity.		2
(d)	High melting point because of: giant structure / lots of bonds / macromolecule		1
	strong bonds / lots of energy to break bonds.		1
	Does not conduct electricity because: there are no free / mobile electrons **or** all its electrons are used in bonding.		1
4 (a)	copper(II) carbonate $\rightarrow$ copper(II) oxide + carbon dioxide	Accept copper carbonate and copper oxide without (II).	1
(b)	12.35 – 7.95 =		1
	4.40 (g)	Accept 4.4 (g)	1
5 (a)	Magnesium loses electrons and oxygen gains electrons.		1
	Two electrons are transferred.		1
(b)	The magnesium ion has a charge of 2+.		1
	The oxide ion has a charge of 2–.		1
6 (a)	It has the same number of electrons and protons.		1
(b) (i)		Only award 1 mark if all three numbers are correct.	1
(ii)	**Protons:** the atomic number / number on the bottom left of symbol		1
	Neutrons: 56 – 26 = 30 / mass number – atomic number		1
	Electrons: 26 – 2 = 24 / atomic number but 2 electrons have been removed to make 2+		1
7	$\dfrac{1.2 \times 10^{-6}}{1.5 \times 10^{-10}} =$	Allow 1 mark for working, even if the answer is not correct.	1
	8000 or 8.0×10^3		1

Table for 6 (b) (i):

Number of protons in the ion	26
Number of neutrons in the ion	30
Number of electrons in the ion	24

Question number	Answer	Notes	Marks
8 (a)	Copper oxide and sulfuric acid		1
(b)	The pH would start high and decrease to below 7.		1
(c)	Nitric acid		1
(d)	M_r CaO = 56 **and** M_r $CaCl_2$ = 111		1
	5.55 g $= \dfrac{1}{20}$ mole or 0.05 mole $CaCl_2$		1
	Reaction ratio: 1:1		1
	$\dfrac{1}{20} \times 56$ or $0.05 \times 56 = 2.8$ g		1
9 (a)	$Fe_2O_3 + 3CO \rightarrow 2Fe + 3CO_2$	Allow 1 mark for: $Fe_2O_3 + CO \rightarrow Fe + CO_2$	2
(b)	Copper is less reactive than carbon / copper is lower in the reactivity series than carbon / or reverse argument (ora).		1
	Aluminium is more reactive than carbon / aluminium is higher in the reactivity series than carbon / ora.		1
(c)	**Cathode:** $Al^{3+} + 3e^- \rightarrow Al$		1
	Anode: $2O^{2-} \rightarrow O_2 + 4e^-$		1

10 (a)

Solid	Start temperature (°C)	End temperature (°C)	Temperature change (°C)		
Ammonium chloride	15	9	–6		
Potassium hydroxide	16	29	+13	Award 1 mark for two correct numbers and 1 mark for two correct signs.	2
Ammonium nitrate	18	4	**–14**		
Sodium hydroxide	17	35	**+18**		

Question number	Answer	Notes	Marks
(b)	Ammonium nitrate		1
	Endothermic reactions cause the temperature to decrease.		1
	Ammonium nitrate had the biggest drop in temperature.		1
11	**Level 3:** Correct descriptions of the development of at least three atomic models and the linking of two scientists.		5–6
	Level 2: Correct descriptions of the development of at least two atomic models and the linking of one scientist.		3–4
	Level 1: One correct description of the development of any atomic model.		1–2
	No relevant content		0
	Indicative content 'Plum pudding' model of the atom / existence of electrons – J.J. Thomson Nuclear model / 'solar system' atom – Marsden and Rutherford Electron orbits – Niels Bohr Existence of neutrons – James Chadwick		
12 (a)	First test – the iodide converted (oxidised) to iodine.		1
	Bromine is more reactive than iodine.		1
	No reaction in second test		1
	because bromine is less reactive than chlorine.		1
(b)		Give 1 mark for the shared pair of electrons.	1
		Give 1 mark if the rest of the diagram is correct.	1

Question number	Answer	Notes	Marks
(c)	Chlorine has weaker forces between molecules / weaker intermolecular forces than iodine. Less energy is needed to separate molecules in chlorine than iodine.	Allow 'stronger' forces for iodine. Allow 'more energy' for iodine.	1 1
(d)	It has no free electrons / the electrons cannot move / all outer electrons are involved in bonding.		1
13 (a)	$\dfrac{14.4}{12}$ = 1.2 mole $1.2 \times 393 = 472\,kJ$	Accept 471.6 kJ	1 1
(b)	Energy is taken in to break (oxygen) bonds. Energy is given out, making (C=O) bonds. More energy is given out than taken in.		1 1 1
(c)	**Any three from:** • Use a more accurate method of measuring the volumes, e.g. a 25 cm³ pipette or a 25 cm³ measuring cylinder. • Use an insulated beaker / polystyrene cup. • Use separate apparatus to measure the acid. • Measure the start temperature before adding the acid. • Wait until there is no further temperature change before measuring the final temperature. • Repeat the method at least two more times.		3

Paper 4: Chemistry 2

Question number	Answer	Notes	Marks
1 (a)	Chlorine		1
(b)	Carbon dioxide		1
2 (a)	Breaking up / decomposing (large molecules)		1
	into smaller / more useful substances / molecules.		1
(b)	**Any one from:** They act as a catalyst. They speed up the reaction. They provide a surface on which the reaction happens.		1
(c)	$C_{16}H_{34} \rightarrow C_{10}H_{22} + 3C_2H_4$	If balancing is not correct, allow 1 mark for: $C_{16}H_{34} \rightarrow C_{10}H_{22} + C_2H_4$	2
3	**Any three from:** Small molecules have weaker (intermolecular) forces / large molecules have stronger forces. Small molecules have lower boiling point / large molecules have higher boiling point. Stronger forces lead to a higher boiling point / weaker forces lead to a lower boiling point. Fractions are separated by different boiling points.		3
4	C_5H_{12}		1
5 (a)	Dylan		1
(b)	The acid was the limiting reagent in her experiment / all of the acid had been used up in her experiment.	Allow: magnesium in excess.	1
	Magnesium was the limiting reagent in the other experiments / all of the magnesium was used up in the other experiments.	Allow: acid in excess.	1
(c) (i)	mean rate = $\dfrac{\text{quantity of product formed}}{\text{time taken}} = \dfrac{94.5}{225} =$		1
	0.42		1
	Unit = cm^3/s		1
(ii)	The volume of gas will double		1
	because the volume of gas is dependent on the amount of magnesium.		1
6 (a)	C and D		1
(b)	$0.86 = \dfrac{\text{distance moved by B}}{7.91} = 0.86 \times 7.91$		1
	$= 6.80$	6.8 / 6.802 / 6.8026 is worth 1 mark.	1
7 (a)			1
(b)	$2C_4H_{10} + 13O_2 \rightarrow 8CO_2 + 10H_2O$	If balancing is not correct, allow 1 mark for: $C_4H_{10} + O_2 \rightarrow CO_2 + H_2O$ If all the numbers are correct, allow multiples for balancing.	2
8	1.69 g	Pure gold = 24 carats, 9 carats = $\dfrac{9}{24}$th gold $\dfrac{9}{24} \times 4.5 = 1.69$ (rounded up)	1

Question number	Answer	Notes	Marks
9 (a)	D, A, C, B	Allow 1 mark if D and A are in the correct order.	2
(b)	It is a simple or complex explanation put forward by scientists		1
	to try and explain observations/facts		1
(c)	It cannot be proved because it was so long ago / there is insufficient evidence to confirm the theory.		1
(d)			1
10 (a)	All six points correctly plotted Smooth curve	Allow 1 mark for four or five points correctly plotted; 1 mark deducted if the graph is plotted dot to dot rather than as a smooth curve.	2 1
(b)	Rate increases as concentration increases.		1
	Increased concentration means more crowded particles / more particles in same space / volume.		1
	More chance of collision / more frequent collisions.		1
(c)	The reaction times would be smaller / shorter / less.		1
11	**Desalination:** removal of salt from the water / changing sea water into drinking water.		1
	Chlorination: adding chlorine to kill bacteria / germs / microorganisms.		1
12 (a)	50%		1
(b)	The yield decreases.		1
(c)	If pressure increases, the reaction goes in the direction that reduces the number of moles of gas / the volume of gases, i.e. the direction with the fewest number of moles of gas.		1
	The equilibrium moves to the right / forward direction / to the product side.		1
(d)	High pressures need energy / equipment that can tolerate high pressures, and this is expensive.		1
	The costs outweigh the benefits of a higher yield.		1

Question number	Answer	Notes	Marks
13 (a)	B Powder reacts faster / line steeper on graph because powder has a larger surface area. More chance of successful collisions / more frequent successful collisions.		1 1 1 1
(b)		The curve must start at reactants, be lower than the original curve and end at products.	1
14	Alloys are harder. Alloys have a lower density. Alloys have a lower melting point. Alloys are stronger.	Allow the reverse argument, e.g. pure metals are softer.	4
15	 Accept curved or square brackets.	Award 1 mark for side links; 1 mark for n; I mark for the rest of the molecule (must have only one CH_3 group).	3
16 (a)	As the reaction progresses, the rate of the backward reaction increases until the rates of the forward and backward reactions are the same.		1 1
(b)	Favours the reverse reaction / equilibrium position moves to the left / moves to the reactant side. Reaction goes in the direction to increase the number of moles / reactants have a larger volume / reaction works to increase pressure on the side with higher moles of gas.		1 1
(c)	No effect on the equilibrium position.		1
(d)	Which reaction direction is exothermic / which reaction direction is endothermic.		1
17 (a)	Nitrogen **and** oxygen from the air react in the high temperatures in the engine.		1 1 1
(b) (i)	**One from:** causes acid rain / causes respiratory problems.		1
(ii)	Remove sulfur from the fuel before use.		1

Paper 5: Physics 1

Question number	Answer	Notes	Marks
1 **(a)**			1
(b)			1
(c)	$V = I \times R = 0.005 \times 1200$ $= 6\ V$ $\dfrac{6\ V}{1.5\ V} = 4$ batteries needed		1 1 1
(d)	Diode Because when it has a positive potential difference the resistance is low. Because when it has a negative potential difference the resistance is high.	Allow low resistance when potential difference is high enough Allow voltage for potential difference in both cases	1 1 1
2 **(a)**		1 mark for each correct column	3

Isotope	Number of protons	Number of neutrons	Number of electrons
Carbon-12	6	6	6
Carbon-13	**6**	**7**	**6**
Carbon-14	**6**	**8**	**6**

Question number	Answer	Notes	Marks
(b)	A neutron changes into a proton by emitting a (high speed) electron.		1 1
(c)	The time taken for the activity / count rate to halve **or** the time taken for the number of nuclei of the isotope to halve.		1
(d)			1

Number of half-lives	Time after death of organism in years	$^{14}C : {}^{12}C$ ratio / 10^{-12}
0	0	1.000
1	5730	0.500
2	11 460	0.250
3	**17 190**	0.125

Question number	Answer	Notes	Marks
(e)		Plotting points correctly Smooth curve	1 1
(f)	10 000 (years)	Accept answers in the range 9500–10 500	1
3 **(a)**	50 Hz, 230 V, ac		1
(b)	It does not need an earth wire.		1

Question number	Answer	Notes	Marks
4 (a)	<table><tr><td>Wire</td><td>Name</td><td>Colour</td></tr><tr><td>A</td><td>**Neutral**</td><td>Blue</td></tr><tr><td>B</td><td>Live</td><td>**Brown**</td></tr><tr><td>C</td><td>**Earth**</td><td>Green and yellow stripes</td></tr></table>	2 marks for all three answers correct, 1 mark for two correct answers	2
(b)	**Any two from:** • The earth wire stops the appliance becoming live. • It carries current if there is a fault. • It stops someone from being electrocuted if there is a fault.		2
(c)	power = potential difference × current P = V × I current = $\frac{960}{230}$ = 4.2(A)	Allow 2 marks for 4.17 or 4.173	1 1 1
5 (a)	Coal		1
(b)	Geothermal		1
(c) (i)	**Any one from:** • causes air pollution • emits greenhouse gases • mines spoil the countryside Accept any other reasonable answer.		1
(ii)	Economic/social reasons – it's cheaper to use fossil fuels. Political reasons – lots of jobs will be lost if we stop using fossil fuels.		1 1
6 (a)	Position 4		1
(b)	Maximum gravitational potential energy at position 1. Changes to kinetic energy in Positions 2–4 (as the cars go down the hill).		1 1
(c)	**Any three from:** The kinetic energy converts back to gravitational potential energy but some energy is lost due to friction / air resistance. Less energy means less height.		3
7 (a) (i)	Missing number is 222	242 – 20 = 222	1
(ii)	Motor C		1
(b)	Electrical power is used to turn around / spin the bit / screwdriver blade. Some power is wasted as heat / sound / vibration.		1 1
(c)	efficiency = $\frac{\text{useful output power}}{\text{total input power}} = \frac{160}{172}$ = 0.93 = 93%	Accept 0.93023 or 93.023%	1 1
8 (a)	A car with a mass of 1400 kg travelling at 12 m/s		1
(b)	work done = force × distance = 24 × 2.5 = 60 J		1 1
(c)	power = $\frac{\text{work done}}{\text{time}} = \frac{60}{4}$ = 15 W	Allow 2 marks for an incorrect answer to 8 (b) correctly divided by 4	1 1
9 (a)		1 mark for drawing appropriate apparatus 1 mark for showing that water is displaced The diagram must have labels to gain full marks	1 1

Question number	Answer	Notes	Marks
(b)	**Any two from:** reduces random experimental error can calculate a mean value can identify anomalous results.		2
(c)	Mass	Accept weight	1
(d)	Nickel		1
10 (a)	Lead		1

(b)	energy = power × time
	
	    = 100 × 85
	
	    = 8500 J
			1
			
			1

(c)	temperature change = $\dfrac{\text{change in thermal energy}}{\text{mass} \times \text{specific heat capacity}}$
	
	    = $\dfrac{9065}{0.5 \times 490}$
	
	    = 37°C
	
	final temperature = 22 + 37 = 59°C
			1
			
			1
			
			1

11	No Energy to melt ice = 0.06 × 334 000 = 20 040(J) 20 040 greater than 16 800 so there is not enough energy available from the water.	Also accept: Energy available in water would melt $\dfrac{16\,800}{334\,000}$ = 0.05 kg or 50 g of ice However, there is 60 g of ice present, so 10 g of ice would not melt.	1 1 1
12	**Level 3:** Correct explanation of why gamma radiation can be used to treat cancer, the risks involved and at least two control measures. **Level 2:** Correct explanation of why gamma radiation can be used and the risks involved, **or** explanation of why gamma radiation can be used and at least two control measures. **Level 1:** Correct explanation that gamma radiation kills cells and at least one risk or control measure. No relevant content **Indicative content** **Gamma radiation** Can penetrate to tumour Is ionising radiation Kills cancer cells Can be focused on particular area Dose can be controlled **Risks** Can kill healthy cells Radiation could cause normal cells to become cancerous and cause further problems **Risk control** Short exposure time Precise targeting of the radiation Calculating the correct dose Balancing risk with likelihood of successful treatment Using shielding on other parts of body		5–6 3–4 1–2 0

Paper 6: Physics 2

Question number		Answer	Notes	Marks
1	(a)	Velocity		1
	(b)	Weight		1
2	(a)	10 m/s		1
	(b)	$acceleration = \dfrac{change\ of\ velocity}{time\ taken}$		1
		$= \dfrac{15-0}{3}$		1
		$= 5$		1
		$= 5\ m/s^2$		1
	(c)	distance 0–3s = 0.5 × 15 × 3 = 22.5 m	distance = area under line	1
		distance 3–6s = 15 × 3 = 45 m	distance = velocity × time	1
		total distance = 22.5 + 45 = 67.5 m		1
	(d)	Slowing down / decelerating / negative acceleration		1
		at a uniform / constant rate.		1
3	(a)	C		1
	(b)	Infrared		1
	(c)	$frequency = \dfrac{wave\ speed}{wavelength} = \dfrac{3 \times 10^8}{0.12}$	1 mark for showing working, 1 mark for the correct answer and 1 mark for the correct unit	1
		$= 2.5 \times 10^9\ Hz$		2
4	(a)	To compare times / speeds in the two different sections.		1
	(b)	If the times are the same, David is correct.		1
		If the time for the car to travel down the second half of the ramp is less than the first, then Gemma is correct.		1
	(c)	Run 2		1
	(d)	$Mean\ time = \dfrac{(0.249 + 0.270 + 0.251)}{3}$		
		$= \dfrac{0.77}{3}$		1
		$= 0.257$		
		$mean\ speed = \dfrac{distance}{time} = \dfrac{1.50}{0.257}$		1
		$= 5.84\ m/s$	Also accept 5.4, 5.836 or 5.8365.	1
5		A and D		1
6	(a)	Force X increases		1
	(b)	Force X = 550 N	Accept weight for force Y	1
		The forces balance each other / the forces reach an equilibrium / force X = force Y	Accept air resistance for force X	1
	(c)	4 and 5		1
	(d)	Parachute has a larger surface area.		1
		Force X therefore increases.	Accept weight for force Y	1
		Force X greater than force Y.	Accept air resistance for force X	1
		Velocity reduces until the forces balance again.		1
7	(a)			1
	(b)	They will move		1
		to point in a circle.		1
8	(a)	Increase current		1
		Increase the number of coils		1
	(b)	Up from the page, towards you		1
	(c)	Wire has a magnetic field only when there is a current.		1
		The wire's magnetic field interacts with the magnet's field.		1

Question number	Answer	Notes	Marks
9 (a)	The reading decreases as the rock is lowered in the water.		1
	No (further) change when the rock is fully submerged.		1
(b)	Use a ruler / measure distance lowered.		1
	Have suitable increments, e.g. measure force every 1 cm lowered		1
(c)	Force Distance lowered		1
10 (a)	Seawater has a higher density than fresh water.		1
	Less seawater would be displaced		1
	to balance the weight of the ship.		1
(b)	distance = speed × time		1
	= 1500 × 0.0270		
	= 40.5 m	To find the distance travelled to the seabed and back, divide by 2	
	$\dfrac{40.5}{2}$		1
	= 20.25 m		1
(c)	Water is more dense than air.		1
	Pressure is proportional to density (if height and gravity are constant).		1
11 (a)	momentum = mass × velocity		1
	= 1200 × 20		
	= 24 000 kg m/s		1
(b)	force = $\dfrac{24\,000}{0.50}$		1
	= 48 000 N		1